International Conference "Development of Eco-Friendly Composite Materials Based on Geopolymer Matrix and Reinforced with Waste Fibers" (FIBER 2019)

IOP Conference Series: Materials Science and Engineering
Volume 706

Montevideo, Uruguay
28 - 29 November 2019

Editor:

Michal Lach

ISBN: 978-1-7138-1036-0
ISSN: 1757-8981

Printed from e-media with permission by:

Curran Associates, Inc.
57 Morehouse Lane
Red Hook, NY 12571

Some format issues inherent in the e-media version may also appear in this print version.

Institute of Physics
Dirac House, Temple Back
Bristol BS1 6BE UK

Phone: 44 1 17 929 7481
Fax: 44 1 17 920 0979

techtracking@iop.org

Additional copies of this publication are available from:

Curran Associates, Inc.
57 Morehouse Lane
Red Hook, NY 12571 USA
Phone: 845-758-0400
Fax: 845-758-2633
Email: curran@proceedings.com
Web: www.proceedings.com

TABLE OF CONTENTS

IOP Conf. Series: Materials Science and Engineering **706** (2019) 011001 doi:10.1088/1757-899X/706/1/011001

Preface

The international conference „Development of eco-friendly composite materials based on geopolymer matrix and reinforced with waste fibers" (FIBER) took place on 28-29 November 2019 in Montevideo, Uruguay. The event was organized by Cracow University of Technology, Poland and Catholic University of Uruguay Damas Antonio Larrañaga, Uruguay as the culmination the great work carried out by 8 universities from 7 countries in the framework of the project supported by the ERANet-LAC 2nd Joint Call (http://www.eranet-lac.eu) within the framework of the grant: "Development of ecofriendly composite materials based on geopolymer matrix and reinforced with waste fibers". Despite the fact that it was the final conference I hope that it will not be the ending, but the beginning of new fruitful cooperation between consortium members as well as the first among the long series the world-recognized conferences organized by common effort European and Latino America countries.

Our goal was to provide a stimulating environment for the presentation and discussion of the most recent results of industrial and academic research carried out in the field of waste management, recycling and urban mining, especially eco-friendly composite materials from waste natural (animal, vegetable or mineral) fibers, for replacing the traditional construction materials, especially those based on the polymer matrix and synthetic fibers, with new ones containing bio-based matrices and fiber derived from waste, for instance making a composite material from wastes from flax, bio-waste straw, nut shells, hen feathers, waste from mineral wool production and other organic waste materials. During the present conference, not only the project results were presented, but also the numerous important research connected with the latest research in the area of advanced eco-friendly materials, waste management, recycling, and urban mining. The conference was an excellent opportunity for all participants to share the important achievements and discuss the current challenges and perspectives in this area.

Many people contributed to the success of the conference, and on behalf of the organizing committee, I would like to thank all the speakers, authors, chairs and reviewers for interesting contributions, providing exciting talks and initiating fruitful discussions. FIBER attracted scientists from 7 countries. After rigorous reviewing, about 20 technical papers were accepted for publication in the conference proceedings.

Eventually, I would like to thank all those who helped to make this event happen. Special thanks go to Prof. Dr. Martín Duarte Guigou and PhD, DSc. Janusz Mikuła, Prof. of CUT for their major contribution to organizing the conference.

PhD Michał Łach

Editor-in-Chief

Faculty of Materials Engineering and Physics

Cracow University of Technology

IOP Conf. Series: Materials Science and Engineering **706** (2019) 011002 doi:10.1088/1757-899X/706/1/011002

Peer review statement

All papers published in this volume of *IOP Conference Series: Materials Science and Engineering* have been peer reviewed through processes administered by the proceedings Editors. Reviews were conducted by expert referees to the professional and scientific standards expected of a proceedings journal published by IOP Publishing.

Environmental and sustainable aspects of green building: A review

F Çiner[1]and N Doğan-Sağlamtimur[1]

[1]Department of Environmental Engineering, Niğde Ömer Halisdemir University, Niğde, 51240, Turkey

E-mail: neslihandogansaglamtimur@gmail.com

Abstract. Public and scientific focus of interest on climate changes, population increase, rapid urbanization is rising day by day. Thus, increased demand for green buildings that reduces negative impacts, and can creates positive impacts on climate, natural environment and sustainable development is a new topic. Green buildings must be not only eco-friendly (efficiently use energy, water, and other resources, use renewable energy, reduce pollution and waste, enable reuse and recycling, have good indoor environmental quality, use of materials that are non-toxic, ethical and sustainable) but also function as the key part to provide a comfortable working/living environment for human activities and improve quality of life. It is realised that green buildings as a worldwide response to raise awareness of the role of human activity in global climate changes, with maximum expression in Leadership in Energy and Environmental Design (LEED) Certification and its five basic requirements: Sustainable Space (SS), Materials and Resources (MR), Energy and Atmosphere (EA), Environmental Quality (EQ), and Water Efficiency (WE).The aim of this study is to review the environmental and sustainable aspects of green buildings. For this goal,environmental and sustainable aspects of green building including its origin, concept, eco-friendly properties, benefits, tools, rating systems, certifications are given and discussed in detail in this study.

1. Introduction

A building as the shelter is designed to provide an indoor environmental quality (IEQ) that is comfortable and attractive/appealing to its users. Having the understanding of the relationship between the IEQ performance in buildings and the overall satisfaction derivable from the environment by the building occupants are necessary variables that will enhance the formulation of the requirements for architectural and building systems design and control. Nowadays, the assessment and advancement of the IEQ in buildings is arousing attention due to its contributions to green buildings and sustainable environment [1-6]. It is therefore relevant to state that providing an excellent indoor environment in a building will contribute substantially to its rating as a green building [3].

The increase in the world population pertains to higher demand for water, energy and natural resources which, in turn, overloads ecosystems and increasingly damages the environment [7]. Sustainable cities necessitate sustainable/green buildings[8].A green building (a green construction or a sustainable building) refers to a structure that uses an environmentally responsible and resource-efficient process throughout a building's lifecycle: from sitting to design, construction, maintenance, operation, destruction and renovation [9].A green building has many different definitions, but it is generally considered to be the planning, design, construction and operation of buildings with utmost economizing of resources (energy, land, water and material), pollution reduction, environmental protection and provision of people with healthy and comfortable interiors [10,11].

Among other manufacturing and production sectors, the construction sector is one of the sectors that should include sustainability and has been the main driving force of sustainable construction and green building movements. There is a consensus in the literature thatgreen buildings perform better than traditional (non-green) buildings in a number of performance/application areas[7].Green building designs are applicableforvarious real estate projects such as: (i) Residential (Housing and Development), (ii) Industrial (Plants, Factories, Industrial Parks,etc.), (iii) Corporate (Office, Research Centres) and (iv) Commercial (Retail: Hotels, Shops, Malls, Showrooms, etc.) [12].Conceptual drawings of a green building are shown in Figure 1[13, 14].

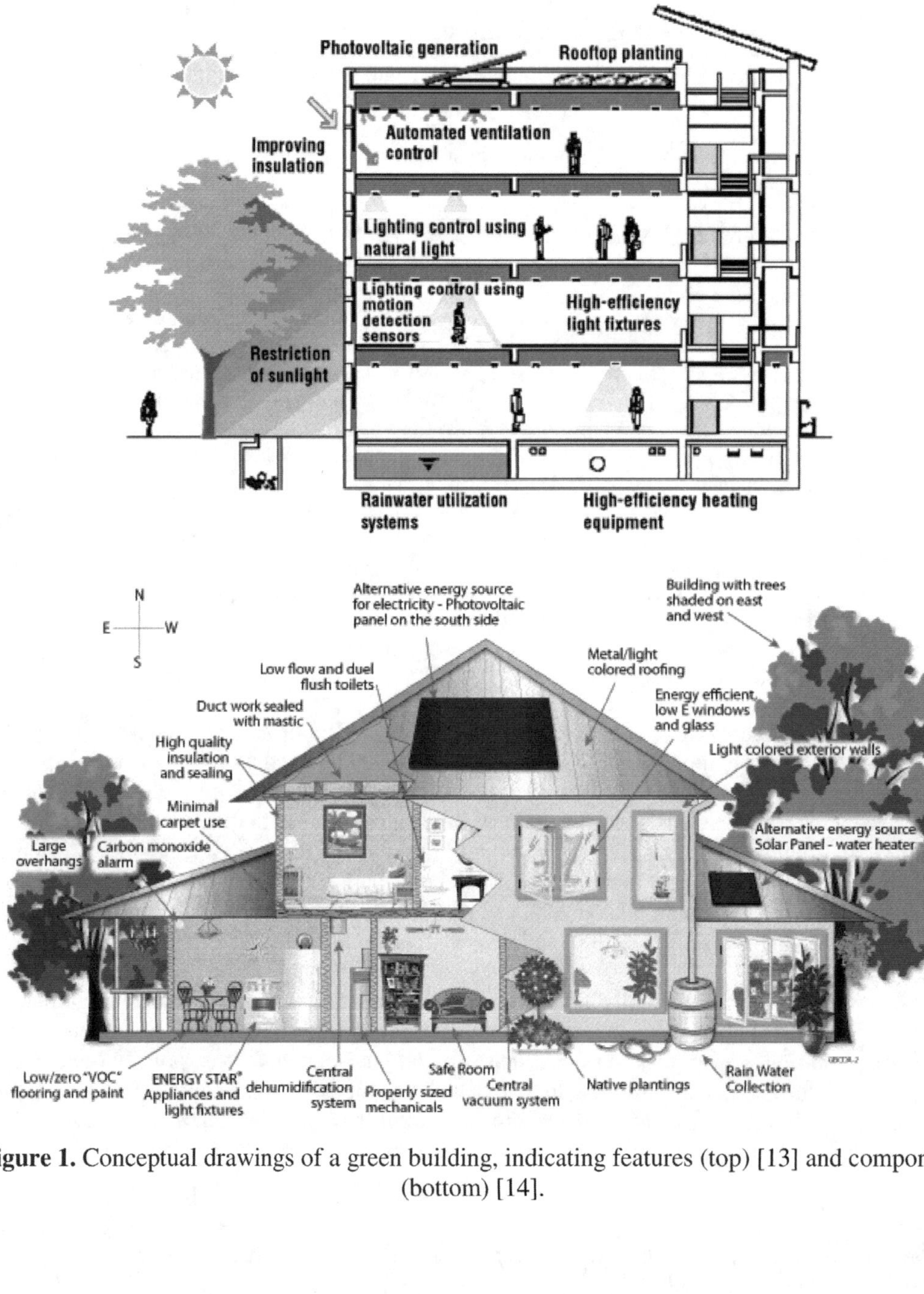

Figure 1. Conceptual drawings of a green building, indicating features (top) [13] and components (bottom) [14].

The aim of this study is to review the environmental and sustainable aspects of green buildings. It also addressesthe issues on its origin, concept, eco-friendly properties, benefits and focuses its tools, rating systems and certifications.

2. Green Building Issues: Its Origin, Concept, Eco-Friendly Properties and Benefits

The history of green buildings dates back to ancient times. Even the cave dwellers utiziled environmentally friendly materials and resided in their homes in compliance with the landscape.Throughout the eras, human interference with the environmental balance causes problems. This was the beginning of the sustainability and green movements.Eventhough the green building movement has been accelerated throughout the past 10 years; its origins can be detected at as early as the end of the nineteenth century [15]. Green building philosophy is derived from the term "Arcology" which stands for a combination of architecture and ecology [16,17]. Green buildings point out an application in the construction sector that prioritize environmentallyresponsible and efficient resource allocation in the whole life-cycleof a building [18].

Green building concept has a worldwide popularity. The negative influence of construction on the environment crucially stimulates the emergence of the green building concept. This concept is a final reaction in addressing environmental and health problems which arise from buildings and minimizing the effects of the construction sector on the natural habitat as well as on individuals [5,19,20]. The concept of green building considers ecology and social conservation in business planning and performance processes, including energy-saving [21].

To eliminate the adverse impacts of buildings on the environment; green buildings, which could create better IEQ for residents with less consumption of natural resources, have begun to develop [11]. Sustainable construction and green buildings are two interchangeable terms that arise from the concept of sustainable development. A green building stands for the reaction of the construction industry towards the sustainability requirements, and thus, besides health and the environment, energy and water efficiency, the reduction of natural resource consumption as the main features of a green building [7,22]. Because of having these properties, green buildings are generally termed as environmentally friendly [23] (eco-friendly) buildings.

Saved energy (30-40%) and water (20-30%), improved indoor air quality, enhanced health and productivity, increased property value, enhanced day light and ventilation, among others, are frequently mentionedefficacies/benefitsrelated to green building [7,12,22,24]. The environmental efficacies of the environmentally friendly building are evident, but there are other stringent grounds for practicing green building tasks that may not be instantly conceivable. Examples are as follows: (i) Healthier and happier employees working in green buildings, reporting improvements in asthma and allergy symptoms and fewer headaches, (ii) Lowered energy costs, (iii) Ability to hold and retain senior talent, (iv) The possibility of selling a green building for more money than a standard building, (v) Additional business opportunities that appeal to the ever-growing conscious consumer pool [25].

3. Green Building Tools, Rating Systems and Certifications

Two main strategies would be utilized to render buildings more sustainable or greener. The first involves the construction of new (green) buildings that corresponds to a key policy in urban sustainability by altering the constructed environment [26]. The second strategy is the sustainable renovation of existing buildings as an alternative to the new structure due to the high investment costs of new buildings [27]. Over the last decade, attention has been paid to the construction of green buildings and to renewing existing buildings with green technologies. The main results of this attention are the introduction of green building councils and the introduction of certification systems to assess the environmental performance of buildings and to document best practices[5].

Since globally recognized performance goals for the green building have not yet been determined, some countries in the world have developed their own tools and systems to assess the performance of green buildings. According to these vehicles and systems known as building environmental assessment methods (BEAM), a building is rated green if it achieves a set of performance targets specified by the accepted green rating system. Since the mid-90s, construction green labels and certificates have emerged in order to meet the demands of the community for sustainable buildings, non-governmental organizations and private companies[7].

The green building provided tremendous development and thousands of buildings were labeled as "green" all over the world. Meanwhile, a number of rating systems or tools have been developed in the world to facilitate the development of green buildings in many countries, such as Leadership in Energy and Environmental Design (LEED) in the United States (US) of America, The Building Research Establishment Environmental Assessment Method (BREEAM) in the United Kingdom, Deutsche Gesellschaft für Nachhaltiges Bauen (DGNB) in Germany, Green Globes in Canada, Green Building Council of Australia Green Star (GS) in Australia, BCA Green MarkScheme in Singapore, Comprehensive Assessment System for Built Environment Efficiency (CASBEE) in Japan,Green Rating for Integrated Habitat Assessment (GRIHA) in India and Building Environmental Assessment Method (BEAM) in Hong Kong andAssessment Standard for Green Building (ASGB) in China [1,2,11, 28].

The American LEED and the British BREEAM can be cited as widely accepted and used green rating systems for green buildings [7].The US Green Building Council (USGBC) classifies and labels green buildings, the so-called green building rating system, which approves a construction area of 1.85 million square meters in more than 150 countries worldwide with LEED, the most widely used green building rating system worldwide[23], as well as the most perfect, effective evaluation system among all green building assessment systems in the world [23,29,30].The LEED certification enables independent verification of the green properties of a building or environment and ensures the design, construction, operation and maintenance of resource-efficient, high-performance, healthy, cost-effective buildings. LEED does not specify the phase to be evaluated. As a result, most buildings focus on the design phase rather than the operational phase [11].The LEED Certification and its five fundamental requirements point to a global response to the green building to maximize awareness of the role of human activity in global climate change: the Sustainable Area (SS); Materials and Sources (MR); Energy and Atmosphere (EA), Environmental Quality (EQ) and Water Efficiency (WE)[8,22].

BREEAM is the world's first building assessment method since its founding in 1990 [16]. BREEAM sets the standard of best practice in sustainable building design, construction and operation and has become one of the most comprehensive and widely accepted measures of environmental performance of a building. BREEAM is widely used in Europe and throughout the world. As the first green building assessment system, it has a profound impact on other green building assessment systems and facilitates the development of green buildings[11].

DGNB is one of the newest and the first in Germany (the most industrial European country and the most active in the construction and development of sustainable cities) [29]. It provides an objective definition and assessment of the sustainability of buildings and city areas. Quality is assessed comprehensively throughout the entire life cycle of the building. The DGNB Certification System can be applied internationally. Because of its flexibility, it can be fully adapted to the various uses of a building and can even meet country-specific requirements. Ecology, economy, socio-cultural aspects, technology, process workflows and quality departments of the field have been certified to fulfill a maximum of 50 sustainability criteria. The system is based on better use of concepts that are common or usual today [25].

Green Globes (GG) is an online green building rating and certification tool used primarily in Canada and the USA. It is interactive, flexible and affordable, and provides market recognition of a building's environmental attributes through third-party assessment. There are many studies that show that GG is similar in terms of criteria with some exceptions (e.g. materials), and is competitive in terms of usability and cost[31].

The Australian Green Building Council (GBCA) launched the green building rating system as "Green Star (GS)". The GS is Australia's only national, voluntary assessment system for buildings and communities. Australia is the largest contributor to greenhouse gases and accounts for 40% of waste; While the GS increases environmental efficiency in buildings, it also aims to increase productivity, create jobs and improve the health and well-being of communities. Initially, the GBCA had green legacy vehicles and by the end of 2015, all these vehicles were replaced by a new set of green building rating tools. These tools include four green building assessment tools for Design, Interiors, Communities, and Performances, and also evaluate the sustainability of projects at all stages of the life cycle. The GS performance tool evaluates green buildings during its operation stage [17, 23].

BCA Green Mark (GM) Scheme was initiated in January 2005 as an attempt to steer Singapore's construction industry towards more environmentally friendly (eco-friendly) buildings. It is aimed to stimulate sustainability in the constructed environment and to promote environmental awareness among developers, designers and builders as they began to conceptualize projects and designing as well as the construction process. It is a benchmarking scheme that includes internationally recognized best practices of environmental design and performance. The BCA GM certification is in favor of the movement of green buildings by encouraging sustainable buildings in the constructed environment in Singapore [32].

CASBEE is an abbreviation that stands for the method for assessing and evaluating the environmental performance of buildings and the constructed environment. It was developed in 2001 by a research committee established by the cooperation of academia, industry and national and local governments, which established the Japan Sustainable Building Consortium (JSBC) under the auspices of the Ministry of Land, Infrastructure, Transport and Tourism. The CASBEE was designed to improve people's quality of life and also to reduce life cycle resource utilization and environmental burdens pertaining to the constructed environment, from a single home to a whole city [33].

GRIHA is own rating systems of India. It is a green building design evaluation system where buildings are rated in a three-tier process. The process initiates with the online submission of documents as per the prescribed criteria followed by on site visit and evaluation of the building by a team of professionals and experts from GRIHA Secretariat. GRIHA rating system consists of 34 criteria categorised in four different sections. Some of them are site selection and site planning, conservation and efficient utilization of resources, building operation and maintenance, and innovation[34].

BEAM is Hong Kong rating tools for green buildings, it is a voluntary private sector initiative and has developed into an internationally recognised suite of rating tools for green buildings that include new buildings, existing buildings and interiors for shops, offices and retail. BEAM Plus rating tool is sub divided into six categories as (i) site aspects, (ii) materials aspects, (iii) energy use, (iv) water use, indoor environmental quality and innovation [35].

The concept of green building was first coined in China in the 1990s, and then the assessment standard was improved. By June 2006, the Ministry of Construction published “the green building assessment criteria” as the first national standard for assessing green buildings in China [11].The ASGB constitutes the basis for environmental rating systems in China, the largest construction market in the world. The ASGB is an assessment instrument for ASGB residence certificate, suitable for all civilian buildings [17].

Green building certification has become a recognition of sustainable, high performance and energy efficient buildings. The agencies that are responsible for the development and administration of green building ratings operate globally and regionally. These agencies facilitate the certification through green building rating schemes, which are voluntary. There are several variants of green building rating schemes such as new construction, core and shell, existing buildings and major refurbishment. Green building rating schemes are used in industry practice as a practical tool for sustainable development[28].

The indicators that have been assigned highest priority in each green building schemeare given in Table 1 along with the relative weight. Green building rating schemes are noticed to differ considerably with respect to the set goals and achievable points. The indicators set forth to measure the “energy performance” and classified under the “energy” category are given the highest priority in 8 out of the 10 green building schemes. However, discrepancies are observed in the share of the total score (or weighted score) allotted to energy indicators. In the DGNB system, highest weight has been assigned to the indicators "lifecycle cost" and "flexibility and adaptability" (9.6% each), which is higher than any other indicator defined in this system including "energy". High priority is provided for indicators "thermal comfort" under the section "environmental quality" and "efficiency in building services" under the section "environmental load" in the CASBEE scheme[28].

Table 1. Indicators with highest priority in green building schemes [17, 28].

Scheme	Indicator with highest priority	Weight	Category
LEED	Optimize energy performance	16.4%	Energy
BREEAM	Reduction of energy use and carbon emissions	10%	Energy
DGNB	Lifecycle cost Flexibility and adaptability	9.6% 9.6%	Cost Economic development
GG	Energy performance	15%	Energy
GS	Efficiency in building services Greenhouse gases emissions	20% 16%	Energy
GM	Building energy	9.2%	Energy
CASBEE	Thermal comfort	14%	Indoor environment
GRIHA	Energy efficiency	13%	Energy
BEAM Plus	Reduction of CO_2 emissions	10.9%	Energy
ASGB	Renewable energy utilization	28%	Energy

4. Conclusions

Green building development is an integral part of the active participation of stakeholders, including government, developers, technology producers and consumers. Social, economic and ecological behaviors of stakeholders can be the key factor in guiding the sustainable development of green buildings.It was stated [36] that green building development can provide promising efficacies, including improvement of ecological environment, ensuring sustainable land use, conservation of ecological systems, facilitating recycling and reuse of materials, improving energy efficiency and minimizing solid waste and CO_2 emissions.

A total of 73 councils in five continents [37] have so far certified many green buildings. According to USGBC and BREEAM sources, the LEED -the most widely used green building rating system in the world- carried out the certification process of numerous green building projects in more than 167 countries. Correspondingly, the BREEAM, the second most widely accepted rating system in the world, certified many green buildings in 77 countries. Studies have indicated that "green" rated buildings perform better than traditional buildings in terms of their sustainability features with IEQ and energy use.

5. Acknowledgments

This study was presented in the 4th International Conference "Innovative Materials, Structures and Technologies" (IMST 2019). This study includes some ideas of the project "Development of eco-friendly composite materials based on geopolymer matrix and reinforced with waste fibers" funded by the European Commission, within the 7th Framework Programme for Research and Technology Development (FP7), Topic #02: Waste management, recycling and urban mining (Project No. ELAC2015/T02-0721) under the ERANet-LAC: Latin America, Caribbean and European Union and supported by a grant of the Turkish Scientific and Technological Research Council of Turkey (TUBITAK). The Turkish project number is 116Y549.

6. References

[1] Darko A and Chan A P C 2016 *J. Habitat Int.* **57** 53-63
[2] Zhang Y, Wang J, Hu F and Wang Y 2017 *Renew. Sust. Energ. Rev.* **68** 262-271
[3] Nimlyat P S 2018 *Build.Environ.* **144** 598-610
[4] Teng J, Mu X, Wang W, Xu C and Liu W 2019 *Sustain. Cities Soc.* **44** 215-226
[5] Balaban O and de Oliveira J A P 2017 *J. Clean. Prod.* **163** S68-S78
[6] Newsham G, Mancini S and Birt B 2009 *Buildings* **41** 897-905
[7] Dwaikat L N and Ali K N 2018 *J. Build. Eng.* **18** 448-453
[8] Sant'Anna D O, Dos Santos P H, Vianna N S and Romero M A 2018 *Sustain. Cities Soc.* **43** 95-110
[9] Vyas S, Ahmet S and Parashar, A 2014 *Int. J. Res.* **1** 23-32
[10] Aysin AS 2011 *Civ. Eng. Environ. Syst.* **28** 231-245
[11] Geng Y, Ji W, Wang Z, Lin B and Zhu Y 2019 *Energ. Buildings* **183** 500-514
[12] Patel C and Chugan P K 2013 *Measuring awareness and preferences of real estate developers for green buildings over conventional buildings:Consumer behaviour and emerging practices in marketing,* Eds. Jayesh Aagja, Ashiwini K. Awasthi and Sanjay Jain, Institute of Management, Nirma University, Himalaya Publishing House, Mumbai, 332-341
[13] Feautures of a green building. In: EcoMENA Echoing Sustainability in MENA. 2019 https://www.ecomena.org/salient-features-of-a-green-building/ Accessed 22 October 2019
[14] What makes a building green?-Green building concept. 2019.https://theconstructor.org/building/what-makes-a-building-green-green-building-concept/7327/Accessed 22 October 2019
[15] Natural Stone Institute. In: Natural Stone Institute. 2019. https://www.naturalstoneinstitute.org/default/assets/File/consumers/historystoneingreenbuilding.pdf. Accessed 22 October 2019
[16] Zhao D, He B, Johnson C and Mou B 2015 *Renew. Sust. Energ. Rev.* **51** 1594-1609
[17] He Y, Kvan T, Liu M and Li B 2018 *Build. Environ.* **133** 19-31
[18] Khoshbakht M, Gou Z, Lu Y, Xie X and Zhang J 2018 *Habitat Int.* **74** 57-65
[19] Tan Y, Shen L and Yao H 2011 *Sci.Habitat Int.* **35** 225-230
[20] Balaban O, 2012 *Habitat Int.* **36** 26-35
[21] Zhukov A, Bobrova Y, Zelenshchikov D, Mustafaev R and Khimich A 2014 *Adv. Mater. Res.* **1025-1026** 1031-34
[22] Yudelson J 2008 *The Green Building Revolution.* Island Press. Washington D.C.
[23] Ding Z, Fan Z, Tam V W Y, Bian Y, Li S, Chethana L M, Illankoon S and Moon S 2018 *Build. Environ.* **133** 32-40
[24] Choi C 2009 *J. Sustain. Real Estate.* **1** 107-138
[25] What is green building? Explaining eco-friendly construction. In: SUNPOWER. 2019.http://businessfeed.sunpower.com/articles/explaining-what-is-green-building.Accessed 22 October 2019
[26] Cidell J and Cope M A 2014 *J. Environ. Plann. Man.* **57** 1763-81
[27] Juan, YK, Gao P and Wang J 2010 *Energ. Buildings* **42** 290-297
[28] Varma C R S and Palaniappan S 2019 *Habitat Int.* **89** 101989
[29] Hamedani A Z and Huber F 2012 *The Sustainable City* **7** 121-132
[30] Sharifi A and Murayama A 2014 *Build. Environ.* **72** 243-258
[31] A practical approach to green building. In: GREENGLOBES. 2019. http://www.greenglobes.comAccessed 22 October 2019
[32] The DGNB system: Global benchmark for sustainability. In: DGNB System. 2019. https://www.dgnb-system.de/en/system/certification_system/index.php. Accessed 22

IOP Conf. Series: Materials Science and Engineering **706** (2019) 012001 doi:10.1088/1757-899X/706/1/012001

October 2019
[33] About BCA Green Mark Scheme. In: Building and Construction Authority. 2019. https://www.bca.gov.sg/greenmark/green_mark_buildings.html. Accessed 22 October 2019
[34] Three primary rating system for green buildings in India. In: Green Clean Guide For the Changing Planet. 2019. https://greencleanguide.com/three-primary-rating-systems-for-green-buildings-in-india/Accessed 22 October 2019
[35] BEAM PLUS. In: Kelcroft. 2019. http://www.kelcroftasia.com/services/beam-plus.html Accessed 22 October 2019
[36] Zuo J and Zhao Z Y 2014 *Renew. Sust. Energ. Rev.* **30** 271-281
[37] Ulubeyli S and Kazanci O 2018 *J. Clean Prod.* **202** 197-212

IOP Conf. Series: Materials Science and Engineering **706** (2019) 012002 doi:10.1088/1757-899X/706/1/012002

Fly ash as a raw material for geopolymerisation - chemical composition and physical properties

K Korniejenko[1], NP Halyag[2] and G Mucsi[2]

[1] Institute of Materials Engineering, Faculty of Materials Engineering and Physics, Cracow University of Technology, Jana Pawła II 37, 31-864 Cracow, Poland.
[2]Institute of Raw Material Preparation and Environmental Processing, University of Miskolc, 3515 Miskolc, Egyetemváros, Hungary.

E-mail: kinga.korniejenko@mech.pk.edu.pl

Abstract. The article discuss the main features of fly ashes as raw material for geopolymerisation. This article is focused on the examination of the fly ash (FA) from the CHP plant in Skawina (Poland) and assessment it as a main component of geopolymers. The article is focused on chemical and physical properties. The characteristic of the FA is presented, including particle size distribution, density, and chemical composition. The all parameters are discussed according to advantages and disadvantages using this FA to geopolymerisation process, including possibilities to achieve better results through proper preparation of the raw material.

1. Introduction

Producing and using coal affects the environment, especially air quality, including emissions from coal combustion such as: carbon dioxide (CO_2), sulphur dioxide (SO_2), nitrogen oxides (NO_x), particulates, mercury and other heavy metals as well as residues created when power plants burn coal - fly ash (FA) and bottom ash. The estimations show that coal is responsible for 45% of energy-related carbon emission[1, 2].Notwithstanding this, coal is the main energy source to produce electricity[3, 4]. Contemporary it is the second largest-energy source worldwide, covering almost 30% of global primaryenergy consumption. Hard coal together with lignite (brown coal) is the leading the world's non-renewable primary energy source with about 40% of globally generated power [2, 5]. Despite the forecaststhat shows the percentage of the coal participation in the word energy sources will be quite rapidly decline, ithold their position. The coal consumption forelectricity generation is almost growing at the same rate as the electricity consumption (2.8% per year versus 3% per year between 2000 and 2017). It is driven primarily by Asian countries such as China and India [6, 7].

The amount of FA growth with the amount of burned coal. The estimations show that solid products of coal combustion such as coarse bottom ash and fine FA are 5–20 wt% of feed coal [1, 8]. For example, a thermal power plant with a capacity of about 1000 MW and with an operation 6600 equivalent hours per year, generates about 69000 tons of a slag and 383 000 tons of a FA [3]. The first one is composed of the residual coal ash that settles on the bottom of the boiler. The second - it is that floats into the exhaust stacks and it is collected by electrostatic precipitators during the coal burning process [1]. Usually, the FA is produced at a temperature between 1200-1700° C from feed coal. In this temperature the minerals in the coal melt to become fluid and on cooling form an amorphous spherical FA particles. The percentage and quality of solid by-products aredepended on various factors, including: type of coal (bituminous, sub bituminous, lignite or anthracite), chemical and mineralogical composition (i.e. containing organic and inorganic materials), percentage of ash in coal, coal particle fineness, combustion technique used, boiler operating conditions such as air/fuel ratio, burners used, type of boiler and process undergone before combustion[6, 8].As a result a FA characteristic is highly variable from source to source or even in different batches from the same source [9, 10].

The FA is a fine, spherical shape, heterogeneous glass-like particles - glassy spheres such as: precipitators, cenospheres and plerospheres. They generally have bumpy surfaces. Moreover, the FV contains glassy fragments, carbon blocks as well as small and spongious grains [1, 4]. The FA has various colour from tan to grey to black. The colour depends mainly on the amount of unburned carbon and presence of iron compounds [4, 7]. The size of particle is usually between 0.01 and 200 μm. Generally, 10% to 30% of particles are larger than 45μm [4,7]. Specific gravity is from 1.9 to 3.0 [4, 9] and the bulk density is between: 1 - 2.5 g/cm^3 [4, 7] and specific surface area is between 2,000 to 6,800 cm^2/g [7].

Taking into consideration mineral composition, the FA is a powdery material, consisting primarily of silicon dioxide (SiO_2) (both amorphous and crystalline), aluminium oxide (Al_2O_3) and iron (III) oxide (Fe_2O_3) and calcium oxide (CaO) [7, 9]. The FA typically contains 90 - 99% inorganic material, 1-9% organic material, and up to 0.5% fluid components [1, 11].The inorganic material is composed of 34 - 80% amorphous materials and 17% - 63% crystalline phases. The crystalline phases are usually: quartz, mullite, and various iron rich phases such as: hematite, goethite and calcite. The FA organic component is characterized by unburned carbon [1, 9].

From the chemical point of view the FA major qualitative compositions are similar to the natural earthy materials [4]. Usually, 95-99% of the FA consists of oxides of Si, Al, Fe, Ca and Ti and about 0.5% to 3.5% consists of Na, P, K, Mg, Mn and S. The rest of the FA is composed of trace elements such as As, B, Ba, Br, Cd, Cl, Co, Cu, Cr, Fe, Ga, Hg, I, In, Mo, Ni, Pb, Po, Rb, Sb, Sc, Se, Sr, Ti, W, V and Zn [4, 12]. It is worth to notice that the FA among theseelements also consists of toxic elements, especially Cr, Pb, Hg, Ni, V, As, and Ba. The concentrations of the toxic trace elements in the FA is from 4 to 10 times higher than in origin coal [1, 4, 7]. Moreover, it may include dioxins and polycyclic aromatic hydrocarbon compounds [4, 13]. The pH of the FA is ranging from 4.5 to 12.0. It mainly depentson particle size and subsequent concentration of trace metals [4].

The most popular classification of the FAs based on ASTM678 standards (ASTM-C618-2). Depending on the chemical and mineralogical composition, there are two types of FAs, namely class F and class C – table 1 [7, 14].

Table 1. Requirements for the chemical composition of class C and F ashes [15, 16].

Type of FA	Content of basic ingredients [%]			
	$SiO_2 + Al_2O_3 + Fe_2O_3$	SO_3	Loss of irrigation	Moisture
Class C	min. 50	max. 5	max. 6	max. 3
Class F	min. 70	max. 5	max. 6	max. 3

The class F is characterised by a high percentage of silica (SiO_2), alumina (Al_2O_3), and iron oxide (Fe_2O_3) – min. 70% and additionally, low percentage of calcium oxide (CaO) – max. 4%, while class C has a high lime percentage (usually contains more than 20%), and constitutes at least 50% ofsilica (SiO_2), alumina (Al_2O_3), and iron oxide (Fe_2O_3) [7, 8].When the FA is used with reactive aggregate for both types of FAs, the so-called alkali content - Na_2O_{eq} (as equivalent Na_2O), which means Na_2O_{eq} = Na_2O + 0,658 K_2O should be max. 1.5% [15].

The FAs-use become more and more popular, but still most FAs are deposited in open dumpswithout any pre-treatment, which involves a high risk ofecosystem pollution with danger to both human health andthe environment [7]. Contemporary, the main directions of the FAs-use are:

construction industry, especially as admixture to concrete and concreteproducts, groutblended cement and feed forclinker, road base/sub-base, ceramic industry, mining application, catalysis, soil modification and stabilization, soil amendment and agricultural use, environmental remediation (against water pollutions), flowable fills, structural fills andembankments, blastig crit and roofing granules, wastestabilization and solidification, oil field services and miscellaneous [1, 13]. The literature also shows the possibility of using the FAs for the production of geopolymers, but the quality of recived material is strongly depend on the chemical composition and physical properties used FA.

2. Material and methods

2.1. Material

The FA was delivered from the coal power plant 'Skawina'(located in: Skawina, Lesser Poland, Poland). The base material is bituminous coal. The power plant generates electricity and heat in its combined production. It has a generating capacity of 492 MW. The plant supplies heat and hot water to the Skawina city and the Western part of Cracow.

This FA is obtained by electrostatic precipitation of fine particles from the exhaust gases from coal-fired furnaces. According to the Polish law [17, 18], the FA is classified as combustion by-product with the no. 10 01 02 (Fly ash from coal).

2.2. Research methods

Physical and chemical characteristic of the FA was given by the supplier. Additional tests have been conducted, including determined particle size distribution of the FA by laser diffraction - Horiba PARTICA LA950 V2 type laser particle size analyser. During the measurement processes to achieve the appropriate dispersity state the distillate water was applied.

3. Results and discussion

3.1. Chemical composition

The table 2 showschemical composition (oxides) of the FA given by the supplier.

Table 2. Chemical composition of the FA from the CHP plant in Skawina (Poland) –supplier declaration [19].

No		[%]	No		[%]
1.	loss on ignition	2.84 ± 0.14	8.	Fe_2O_3	7.60
2.	SO_3	0.95 ± 0.24	9.	$SiO_2 + Al_2O_3 + Fe_2O_3$	78.21 ± 1.28
3.	chloride (Cl)	0.034 ± 0.010	10.	MgO	3.06 ± 0.23
4.	CaO	0.02 + 0.01	11.	P_2O_5	0.0008 (8 ± 1 mg/kg)
5.	SiO_2 (reactive)	35.86 ± 0.64	12.	Na_2O	1.72
6.	SiO_2	47.81	13.	K_2O	4.62
7.	Al_2O_3	22.80	14.	Na_2O_{eq}	4.76 ± 0.47

The results of the analyseof the FA according to supplier's declaration show useful the FA for the production of geopolymer materials. The FA is rich in oxides such as SiO_2 and Al_2O_3 and contains low amount of CaO.According to ASTM-C618-2, it belongs to class F.

The chemical composition of the FA as a raw material is very important for the formation of bonds characteristic for geopolymers. In the literature, the terms geopolymers and alkaline activated materials are very often used as a synonyms. However, this approach is not correct due to the differences in the structure of these materials [20, 21].

Geopolymers consist of co-polymers of silicon and aluminium, which are stabilized by metal cations (M^+), most often sodium, potassium, lithium or calcium, and bound water [20, 22]. In addition to these chains, the material usually contains other phases, i.e. silicon oxide, unreacted aluminosilicate substrate and zeolites [20, 23]. The geopolymer is characterized by the presence of SiO_4 and AlO_4 tetrahedrons in a three-dimensional structure. They are alternately bound by oxygen atoms. Binding usually occurs in strongly alkaline aqueous solutions, but the reaction can also be carried out in acids [20, 24]. The reactive aluminosilicates are dissolved, and next in the polycondensation process the tetrahedrical structures $[SiO_4]_{4-}$, $[AlO_4]_{5-}$ combine together with the corners, forming amorphous or sub-crystalline spatial aluminosilicate structures [21, 25]. The presence of monovalent Na^+, K^+ or other metal cations balances the negative charge of this structure [20, 21]. Considering the entire geopolymerization process, alkaline activation is only the first step in creating geopolymer materials [20, 25]. In particular, differences in structure are visible in NMR microstructural studies, but it is possible to predict the structure of the material based on the composition of the raw material used for production [21].

Alkaline activated materials do not form 3D networks, but only a 2D structure. This affects their properties. The different structures mean that geopolymer and alkaline activated materials have different physico-chemical and functional properties, especially resistance to chemical agents, functional properties such as fire resistance and long-term properties - durability [20]. Wherein, the mechanical properties of alkaline activated materials could be even higher than that of geopolymers, especially in the short term [20].

In this case, the change in Si:Al ratio and distribution of essential oxides for geopolymerisation (SiO_2 and Al_2O_3)are important. It significantly influences on the position of most intensive band and finally affects final strength of geopolymers[26, 27].Moreover, weight ratios such as: SiO_2/Al_2O_3, CaO/Al_2O_3, CaO/SiO_2, SiO_2/Na2O and Al_2O_3/Na_2O had very scattered effect on mechanical properties [27, 28]. For example, previous studies [9] have demonstrated that the FA with a high reactive Si:Al ratio ($\geq$5) was sodium aluminate activated and produced geopolymers with low to moderate 'as cured' compressive strengths. Despite the low mechanical properties, final materialhad excellentdimensional stability during heating and greater compressivestrengths after heating. In contrast,the FA with a low reactive Si:Al ratio had high'as cured' compressive strengths but poor dimensional stabilityduring heating and reduced compressive strengths afterheating [9].It is hard to specify exactly amount/ values of individual elements or their oxides in the base material, because it is one of the many factors that could influence on the whole process.

The chemical composition of the raw materialis particularly important for material structure forming. The high aluminium content and low calcium content promote the formation of geopolymer materials – 3D structure. Too high CaO content means that the binding process is too fast and thus the geopolymerisation process is not carried out properly. The polycondensation, that is main binding mechanism in geopolymers, needs time. Quick binding may cause insufficient dissolution of active ingredients in the mass and then an alkaline activated material with a slightly different structure than geopolymers is created[24, 29].

The other compounds included in FAs, which may also have a negative effect on the geopolymerisation process, are: high content of sulphur compounds, unburned coal or some iron compounds [25]. The sulphur compounds (SO_3) could lead to instability in volume, which in turn has an impact on durability of final material [7]. Other sulphur compounds could also accelerate activation [7, 30]. The high content of unburned carbon increases the water requirement. Additionally, carbon particles have a very strong affinity and attraction to the organic chemical admixtures,thesealso influence on final properties, including the hardeningprocess. In this case not only amount of it has significance, but also surface area, and type of carbon in terms of its polarity and particles size [7, 31]. With an increase in the content of loss on ignition (LOI), the content of other components (especially SiO_2)decrease. It affects for the activity of theFA - reducing the reactivity of pozzolanic constituents.Additionally, as the quantity of unburned carbon particles increases, the mechanical properties of geopolymers decrease, in particular a compressive strength [7, 25]. A large amount of iron (on the order of several percent) in the form of hematite or magnetite could negatively affect the dissolving process of ash grains, which makes the geopolymer formation difficult. These compounds usually form on the surface of ash grains the layer and hinder the access of the liquid phase to its vitreous [25, 32]. However, some research shows that even high content iron oxide (Fe_2O_3) should positively influence on 3D structure [7, 33] and could be important in the high temperature applications [34].

Taking into consideration chemical composition, the best for theproduction of the geopolimer that has optimal properties, seems to be the FA class F that contains: less than 5% of unburnedmaterial, max. 10% of Fe_2O_3 and lower amount of CaOcontent [7, 35]. This kind of characteristic has the FA from the coal power plant 'Skawina'. However, it should be emphasized that the chemical composition of the FA could be changeable. Other requirements is the high value of reactive silica. According to the literature the best will be reactive silica amount between 40 to 50% [35, 36]. The FA from 'Skawina' has smaller amount than optimal. However, there aresome possibilities for increasing the reactive level ofthe FA, exemplary by milling [37] or supplementation by proper admixtures [38].

The other important factor, taking into consideration potential applications, is the content of heavy metals and toxic elements [39, 40]. The FA from the coal power plant 'Skawina' has limits for this kind of elements much below standard.

3.2. Physical properties

The physical properties also are important for the final material's properties. The FA from the coal power plant 'Skawina' has following physical properties: fineness of 16.7%, density 2.22 g /cm^3 and an index of pozzulan activity after 28 days at level 92.0%, and after 90 days at 108.8% [41]. The size of particles was investigated. The median size was 14.54 μm. The mean size was 23.88 μm.About 88% of all particles werebelow 45 μm. The detailed results are presented infigure 1.

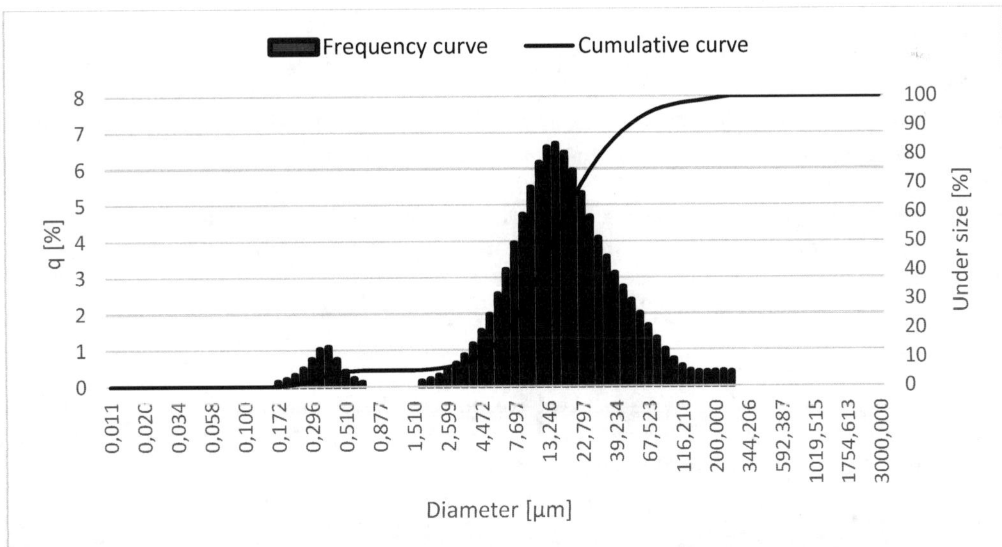

Figure 1. The particle size distribution of the FA.

The particlesize distribution ofthe FA plays an essential role in the development of the mechanical strength of the obtained materials and it is connected with the reactivity of the FA [32, 42]. Contemporary, according to the literature sources, the most important are two mechanisms. Firstly, when the particles are up of 45 μm, this fact has a negative influence on the water requirement.

Particles sizes have also an important effect on the reaction rate of the FA at early stages. Secondly, once diffusion and dissolution of materials occur in concentrated pastes, surface the area of the particles could play an important role in determining the kinetics of different processes [7, 27]. The research shows when the particle fraction sized higher than 45 μm is removed, the mechanical strengths increase, reaching even 70 MPa compressivestrength after 1 day [32, 34, 43].

Other important role plays fineness and specific surface area. The mechanical properties of geopolymers increase together with increase the specific gravity, decrease offineness and increase specific surface area [34].

The FA from the coal power plant 'Skawina' has good physical parameters for geopolymerisation process. The content of particle smaller than 45 μm is about 88%. According to the literature, the good FA should have between 80 to 90% of particles smaller or in the range of 45 μm [7].

4. Conclusions

The analyses confirm the usefulness of the FA from the CHP plant in Skawina (Poland) for geopolimerisation process. The chemical composition of this FA is typically for class F. It contains less than 5% of unburned material and less than 10% of iron compounds. Additionally, it has lower amount calcium content. The amount of the reactive silica in investigatedmaterial is reasonable – ca. 36%, however a little bit under optimal value (40 to 50%). This FA has also good physical parameters for the geopolymerisation process. The content of particle under the size 45 μm is ca. 88%. According to the literature, it is optimal value, also fineness declared by supplier shows that it could be useful material for application in geopolymers production.

This paper analyses only chemical and physical properties. To receive full information additional mineralogical and morphology analysis are required. The further characteristic of the FA from the CHP plant in Skawina (Poland), including the morphology examined by Scanning Electron Microscopy (SEM), the structure investigated by Fourier Transform Infrared Spectroscopy (FTIR) and X-ray diffraction (XRD) is a topic of the second article.

Acknowledgements

1. The research was made partly by the support of Tempus Public Foundation in the framework Bilateral State Scholarships.
2. This work has been financed by the by Polish National Agency for Academic Exchange under the academic exchange programme as part of bilateral cooperation with Hungary.
3. This work has been supported by the ERANet-LAC 2nd Joint Call (http://www.eranet-lac.eu) funded by the National Centre for Research and Development in Poland, within the framework of the grant: *Development of eco-friendly composite materials based on geopolymer matrix and reinforced with waste fibers*.

References

[1] Belviso C 2018 *Prog. Energ. Combust.* **65** 109
[2] Word Energy Council 2016 World Energy Resources | 2016 (London: World Energy Council)
[3] Eliche-Quesada D, Sandalio-Pérez J A, Martínez-Martínez S, Pérez-Villarejo L and Sánchez-Soto P J 2018 *Ceram. Int.* **44** 4400
[4] Yadav V K and Fulekar M H 2018 *Int. J. of Advance Eng.& Research Develop.* **5** (04) 768
[5] Dai S and Finkelman R B 2018 *Int. J. Coal Geol.* **186** 155
[6] Sebi C and Lapillonne B 2019 *The Conversation*, published on line: https://theconversation.com/explaining-the-increase-in-coal-consumption-worldwide-111045 (access: 19-09-2019)
[7] Bouaissi A, Li L Y, Moga L M, Sandu I G, Al Bakri Abdullah M M and Sandu A V 2018 *Rev. Chim. (Bucharest)* **69** (7) 1661
[8] Guzmán-Carrillo H R, Pérez J M, Aguilar Reyes E A and Romero M 2018 *Int. J. Environ. Sci. Technol.* **15** 1757

[9] Rickard W D A, Temuujin J and van Riessen A 2012 J. *Non-Cryst. Solids* **358** 1830
[10] Oh J E, Jun Y and Jeong Y 2014 *Cement Concrete Comp.***50** 16
[11] Vassilev S V, Vassileva C G 2005 *Energ. Fuels* **19** 1084
[12] Roy W R, Thiery R G, Schuller R M and Suloway J J 1981 *Coal fly ash: a review of the literature and proposed classification system with emphasis on environmental impact* (Champaigm, Illinois: Environ. Geol. Notes 96)
[13] Zhuang X Y, Chen L, Komarneni S, Zhou C H, Tong D S, Yang H M, Yu W H and Wang H 2016 *J. Clean. Prod.* **125** 253
[14] Risdanareni P, Puspitasari P and Januarti Jaya E 2017 *MATEC Web of Conferences* **97** 01031
[15] Długosz P, Darłak P, Drzymała P and Sobczak J J 2012 *Prace Instytutu Odlewnictwa* **LII** (1) 5
[16] Zapotoczna-Sytek G, Łaskawiec K, Gębarowski P, Małolepszy J and Szymczak J 2013 *Popioły lotne nowej generacji do produkcji autoklawizowanego betonu komórkowego* (Katowice: Wydawnictwo Instytut Śląski)
[17] Act of 14 December 2012 on waste, J. of Laws of 2013, item 21, with later changes
[18] Regulation of the Minister of the Environment of December 9, 2014 regarding the waste catalog, J. of Laws 2014 item 1923
[19] Łach M, Korniejenko K and Mikuła J 2016 *Procedia Engineer.* **151** 410
[20] Davidovits J 2015 *Geopolymer Chemistry and Applications* (Saint-Quentin, France: Geopolymer Institute)
[21] Provis J L and van Deventer J 2014 *Alkali Activated Materials. RILEM State-of-the-Art Reports*, (Dordrecht: Springer)
[22] Król M and Błaszczyński T Z 2013 *Izolacje* **5/2013** 38
[23] Rajczyk K, Giergiczny E and Szota M 2015 *Prace Instytutu Ceramiki i Materiałów Budowlanych* **23** 79
[24] Yang T, Han E, Wang X and Wu D 2017 *Appl. Surf. Sci.* **416** 200
[25] Mikuła J 2014 *Rozwiązania proekologiczne w zakresie produkcji. Nowoczesne materiały kompozytowe przyjazne środowisku* (Kraków: Wydawnictwo Politechniki Krakowskiej)
[26] Koleżyński A, Król M and Żychowicz M 2018 *J. Mol. Struct.* **1163** 465
[27] Tennakoon C, Nazari A, Sanjayan JG and Sagoe-Crentsil K 2014 Construction and Building Materials **71** 72
[28] Singh NB 2018 *Minerals***8** 299
[29] Dombrowski K, Buchwald A and Weil M 2007 *J. Mater. Sci.* **42** 3033
[30] Gorhan G and Kurklu G 2014 *Compos. Part B-Eng.* **58** 37
[31] Jayant D 2013 Mineral Admixtures in Cement and Concrete (Boca Raton, Florida: Taylor & Francis Group)
[32] Fernández-Jiménez A and Palomo A 2003 *Fuel* **82** 2259
[33] Ken P W, Ramli M and Ban CC 2015 *Constr. Build. Mater.***77** 370
[34] Rickard W D A, van Riessen A and Walls P 2010 *Int. J. Appl. Ceram. Technol.* **7** 81
[35] Brahammaji S V and Muthyalu P V 2015 *Int. J. Adv. Eng. Technol.* **8** (4) 574
[36] Nurwidayati R, Ulum M B, Ekaputri J J, Triwulan M and Suprobo P 2016 *Mater. Sci. Forum* **841** 118
[37] Mucsi G, Kumar S, Csőke B, Kumar R, Molnár Z, Rácz A, Mádai F and Debreczeni Á 2015 *Int. J. Miner. Process.* **143** 50
[38] Łach M, Korniejenko K, Hebdowska-Krupa M and Mikuła J 2018 *MATEC Web of Conferences* **163** 06005
[39] Mierzwiński D, Łach M, Mikuła J and Goły M 2018 *E3S Web of Conferences* **44** 00110
[40] Wang X, Li A and Zhang Z 2016 *Procedia Environ Sci* **31** 440
[41] Mikuła J and Korniejenko K 2015 Innovative, Cost-Effective and Eco-friendly Fibre-based Materials for the Construction Industry (Kraków: Wydawnictwo Politechniki Krakowskiej)
[42] Komljenovic M, Baarevic Z and Bradic V 2010 *J. Hazard. Mater.* **181** (1-3) 35
[43] Somna K, Jaturapitakkul C, Kajitvichyanukul P and Chindaprasirt P 2011 *Fuel* **90** 2118

IOP Conf. Series: Materials Science and Engineering **706** (2019) 012003 doi:10.1088/1757-899X/706/1/012003

Building panel production incorporated with fly ash

G Yıldırım[1], H Keser[1], N Doğan-Sağlamtimur[1,2] and F Çelik[3]
[1]Department of Environmental Engineering, Niğde Ömer Halisdemir University, Turkey
[2]Nanoteknoloji Application and Research Center, Niğde Ömer Halisdemir University, Turkey
[3]Department of Civil Engineering, Niğde Ömer Halisdemir University, Turkey

E-mail: neslihandogansaglamtimur@gmail.com, nds@ohu.edu.tr

Abstract. With the increase in industrialization, the amount of fly ash (FA), which is one of the industrial solid wastes causing environmental and human health problems, is increasing day by day. Waste valorization and reuse methods need to be developed for this waste ash. The construction industry is the main sector where it is used. In this study, it is aimed to evaluate the usability of industrial FA in the production of economical and useful building panels and to determine their physical-mechanical properties. FA supplied from İsken Sugözü Thermal Power Plant (ISTPP), ordinary Portland cement (OPC) supplied from ÇİMSA Cement Industry and Trade Inc., molding plaster (MP) and lime are the materials of this study. Particle distribution curves were determined by using laser scattering method for FA_{ISTPP}. In this study, compressive and flexural strengths, water absorption, bulk density and porosity as mechanical and physical tests were made in the building panels having the combination of FA_{ISTPP}+cement FA_{ISTPP}+MP and FA_{ISTPP}+lime in contribution ratios of 10, 30, 50, 70 and 90%FA_{ISTPP}. The results of the experiments show that the highest compressive strengths values in 7 and 28 days found in the building panels having combination of 90% cement+10% FA_{ISTPP} are 57.30 and 64.20 MPa, respectively. It was determined that the value-added building panels could be produced from FA_{ISTPP}, an industrial solid waste, in the appropriate proportions of cement contribution.

1. Introduction

Fly ash (FA) that is generated during the combustion of coal for energy production is an industrial by-product recognized as an environmental pollutant. Because of the environmental problems caused by the FA, considerable research has been undertaken on the subject worldwide. It has been found that FA is a promising adsorbent for the removal of various pollutants. The adsorption capacity of FA may be increased after chemical and physical activation. It was also found that FA has good potential for usage in the construction industry [1-5]. Finding means of utilizing waste products is a very important field of research at the moment.

Cement is made by heating a mixture of limestone and clay, or other materials of similar bulk composition and sufficient reactivity, ultimately to temperature of about 1450 °C. Particular fusion occurs, and nodules of clinker are produced. The clinker is mixed with a few percent of calcium sulfate and finely ground, to make the cement [6]. Cement, one of the most important building materials, is a binding agent that sets and hardens to adhere to building units such as stones, bricks, tiles, etc. Cement generally refers to a very fine powdery substance chiefly made up of limestone (calcium), sand or clay (silicon), bauxite (aluminum) and iron ore, and may include shells, chalk, marl, shale, clay, blast furnace slag, slate [7-9]. Cement mixed with water causes a chemical reaction (hydration process) and forms a paste that sets and hardens to bind individual structures of building materials. Cement is an integral part of the urban infrastructure. It is used to make concrete as well as mortar, and to secure the infrastructure by binding the building blocks. Concrete is made of cement, water, sand, and gravel mixed in definite proportions, whereas mortar consists of cement, water, and lime aggregate [10, 11]. These are both used to bind rocks, stones, bricks, and other building units, fill or seal any gaps, and to make decorative patterns. Cement, though different from the refined product found nowadays, has been used in many forms since the advent of human civilization. The cement used today has undergone experimentation, testing and significant improvements to meet the needs of the present world, such as developing strong concretes for roads and highways, hydraulic mortars that endure sea water and stucco for wet climates [12-15].

Molding plaster/Gypsum plaster (calcium sulfate hemihydrate) is a material largely employed in civil construction. It is produced by the calcination of the mineral gypsum (calcium sulfate dihydrate) in low temperatures related to other binders. Molding Plaster (MP), commonly referred to as "Plaster of Paris", is a beta-hemihydrate gypsum product manufactured from high purity natural occurring white gypsum rock. It is commonly used as a good white base material for formulating various interior oriented construction and industrial compounds, for simple 3-D statuary, architectural castings, and temporary templates where strength and hardness are not important [16]. The molding plaster can be considered as a low energy environmental-friendly binder [17]. The calcination temperature for molding plaster production is low, in the range of 125-180°C, it loses water vapor and the carbon dioxide emissions come from the kiln heat [18]. The use of lime in plastering applications dates back to the construction of the Egyptian pyramids about 4000 BC. Plastering methods were refined by the Greeks and Romans. Early plaster systems could contain lime, gypsum, marble dust, fibers and sand in up to three coats. Modern plasters can contain Portland cement or gypsum products. Exterior plasters, often called stucco, normally contain Portland cement because of its low solubility in water. Interior plasters usually contain gypsum [19].

Lime is a truly versatile material widely used in the construction of buildings. It can be utilized in masonry applications as a component of mortar or of the masonry unit. Lime has been used in building construction for thousands of years to create durable mortar and plaster. Lime provides some benefits for mortar and plaster in both the plastic and hardened state. In the plastic state, lime can enhance workability and water retention. In the hardened state, lime products react with carbon dioxide to regenerate calcium carbonate or limestone. This is a slow, gradual process that increases the hardness of the finished surface and allows for the closing of hairline cracks by a process called autogenous healing. Since initial strength is needed in most applications, additives such as gypsum, cement or "pozzolans" are mixed with lime in construction applications. Lime can react with pozzolanic materials in the mortar or plaster to produce a cement-like product. The strength of lime-based mixes can be modified according to the needs of the particular application [20]. Until the advent of CEMI in the 1800's, hydraulic lime was the principal binder for use in construction [21]. Both hydraulic lime and CEM I are synthetic materials manufactured by the thermal decomposition of a source of calcium carbonate (typically limestone) at high temperatures [22].

In this study, the usability of cement, molding plaster (MP) and lime incorporated with FA which is supplied from İsken Sugözü Thermal Power Plant (ISTPP)-Adana, Turkey for the production of building panel was investigated.

2. Material and Methods

2.1. Material

FA supplied from ISTPP (FA_{ISTPP}), CEM I 42,5 R type Portland cement supplied from ÇİMSA Cement Industry and Trade Inc., MP and lime in commercial types are the materials of this study.

No tests were performed to determine the chemical properties of the cement, the data of the manufacturer/supplier was used.

Table 1. Chemical composition of FA_{ISTPP}, cement, MP and lime.

Elements	Chemical Composition Results (%)			
	FA_{ISTPP}	Binders		
		Cement	MP	Lime
SiO_2	62.28	3.59	0.40	-
Al_2O_3	21.46	18.9	0.10	-
Fe_2O_3	7.01	3.36	0.11	-
CaO	1.53	63.6	40.1	-
MgO	2.37	1.57	0.22	3.00
SO_3	0.07	2.65	49.1	2.01
K_2O	3.81	0.40	0.02	-
Na_2O	0.26	0.77	-	-
Others	1.21	1.57	-	-
H_2O	-	-	-	2.00
Loss on ignition (LOI)	1.78	3.59	9.95	6.00

2.2. Method

In the study, preliminary experiments were carried out according to the chemical composition of the FA_{ISTPP} (Table 1). Sieve analysis (TS EN 13279-1 standard, Figure 1), combustible material (Table 2) and tight bulk density tests (Table 3) were performed for the FA_{ISTPP}. Then, the experiments were done to produce building panel specimens (Figure 2) incorporated with FA_{ISTPP}. Sample sets (FA_{ISTPP}+cement, FA_{ISTPP}+MP, FA_{ISTPP}+lime) were subjected to physical and mechanical tests in the second stage. These tests are (i) flexural strength (FS), (ii) compressive strength (CS), (iii) bulk density (BD), (iv) porosity and (v) water absorption (WA).

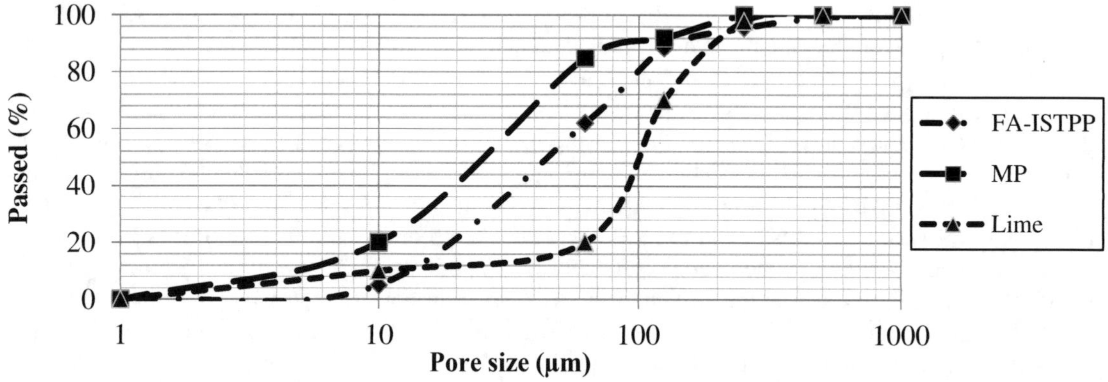

Figure 1. Sieve analysis of FA_{ISTPP}, MP and lime.

Figure 2. Building panel specimens produced in this study.

Table 2. Combustible materials of the FA_{ISTPP}, MP and lime.

Materials	Temperature (°C)	Waiting time (h)	Sample Weight (g)	Combustible Material (%)
FA_{ISTPP}	750	1	20 g	7.5
MP	750	1	20 g	6.3
Lime	750	1	20 g	8.2

Table 3. The tight BD values of FA_{ISTPP}, cement, MP and lime.

	FA_{ISTPP}	CEMI	MP	Lime
BD (g/cm^3)	1.20	1.36	1.01	1.30

Since the BDs of FA_{ISTPP}, cement, MP and lime used in the study were different from each other, weight basis was used to determine the mixture parameters. In the mixture calculation, TS EN 520+A1, TS EN 3234, TS EN 14190, TS EN 13950, TS EN 12859, TS EN 13279-1 and TS EN 13279-2 standards were used. In this study, various combinations were made by using FA_{ISTPP}, cement, MP and lime mixtures (FA_{ISTPP}+cement, FA_{ISTPP}+MP, FA_{ISTPP}+lime). Preliminary experiments were performed. According to TS EN 13279-2 standard, flow test was applied. According to TS EN 12390-5 standards, building panel specimens in 4x4x16 cm were produced.

The curing process was applied to all samples in fully saturated condition in water for 7 and 28 days. After curing time, CS tests were performed for FA_{ISTPP}+cement building panels. And more, the specimens prepared by using FA_{ISTPP}+MP were cured firstly at the room conditions and then they were also cured at 40 °C in oven after 7 days for 24 hours. On the other hand, the specimens produced by mixing FA_{ISTPP}+lime were cured in room condition for 7 and 28 days. After that, CS tests were performed when they reached to constant weight. Porosity was determined for FA_{ISTPP}+cement building panels but could not performed in FA_{ISTPP}+MP and FA_{ISTPP}+lime specimens since water caused them poured.

FS and CS of building panels in 40x40x160 mm size prismatic specimens were tested according to ASTM C 348and ASTM C349at 7 and 28 days, respectively. The BD, WA and porosity of these panels were determined with respect to ASTM C642 at 28 days.

3. Results

In this study, with contribution of FA_{ISTPP}, building panels in the mixture of 10-100% of cement, MP and lime were produced. Mechanical and physical tests (FS, CS, BD, porosity and WA) of these panels were made (Table 4 and 5).

FS values of the cemented building panels incorporated with FA_{ISTPP} at 7-day and 28-day were determined the lowest in 10% cement+90% mixture (0.58 and 0.93 MPa), and the highest in 90% cement+10% FA_{ISTPP} mixture (10.2 and 15.1 MPa). In these building panels, the lowest CS values at 7 and 28 days (1.90 and 4.15 MPa) were found in 10% cement+90% FA_{ISTPP} mixture, whereas the highest (57.3 and 64.2 MPa) values were determined in 90% cement+10% FA_{ISTPP} mixture (Table 4).

In FA_{ISTPP}+MP building panels, the lowest and highest values (0 and 16.4 MPa) were determined in 90% MP+10%FA_{ISTPP} and 90% MP+10% FA_{ISTPP} combinations, respectively. In these building panels, the lowest CS value (0.99 MPa) for 7 days was determined in 50% MP+50% FA_{ISTPP} mixture and the highest CS value (6.80 MPa) was found in 90% MP+10% FA_{ISTPP} mixture (Table 4).

The FA_{ISTPP}+lime building panels were dissolved in 7 days. The lowest and highest FS values (0.42 and 0.87 MPa) were determined in the mixture of 70% lime+30% FA_{ISTPP} and 90% lime+10% FA_{ISTPP}, respectively, at 28-day. The lowest 7-day CS value (1.14 MPa) was found in 50% FA_{ISTPP}+50% lime mixture, and the highest CS value (1.12 MPa) was found in 70% lime+30% FA_{ISTPP} mixture. For 28-day, the lowest CS value (0.43 MPa) was determined in 70% FA_{ISTPP}+30% lime mixture, and the highest CS value (1.83 MPa) was determined in 90% lime+10% FA_{ISTPP} mixture (Table 4).

Table 4. FS ve CS test results of building panels produced by using FA_{ISTPP}+cement, FA_{ISTPP}+MP and FA_{ISTPP}+lime mixtures.

CEM1 (%)	FS		CS		MP (%)	FS		CS		Lime (%)	FS		CS	
	7-day	28-day	7-day	28-day		7-day	28-day	7-day	28-day		7-day	28-day	7-day	28-day
10	0.58	0.93	1.90	4.15	10	-	0	-	0	10	0	0.50	0	1.39
30	3.10	6.80	10.60	25.30	30	-	0	-	1.30	30	0	0.48	0	0.43
50	7.20	9.80	21.60	39.90	50	-	0.99	-	4.57	50	0	0.42	1.14	0.88
70	7.57	12.60	34.80	46.10	70	-	3.91	-	8.36	70	0	0.42	1.12	1.60
90	10.20	15.10	57.30	64.20	90	-	6.80	-	16.40	90	1.25	0.87	1.62	1.83
100	12.53	18.03	67.43	75.42	100	-	8.13	-	19.68	100	-	-	-	-

In FA_{ISTPP}+cement combinations, BD values were found as 1.95 and 2.04 g/cm^3 for building panel produced using 90% cement+10% FA_{ISTPP} mixture and 100% cement. The mixture with the highest porosity value (42.3%) was 90% cement+10% FA_{ISTPP}. The porosity value of building panel produced from 100% cement was 37.4%. The highest WA value (39.3%) was measured for FA_{ISTPP} mixtures in 10% cement+90% FA_{ISTPP} ratio. In the building panel produced from 100% cement, the WA value was measured as 26.1% (Table 5).

Table 5. BD, porosity and WA test results of building panels produced by using FA_{ISTPP}+cement, FA_{ISTPP}+MP and FA_{ISTPP}+lime mixtures.

CEMI	BD (g/cm^3)	Porosity (%)	WA (%)	MP	BD (g/cm^3)	Porosity (%)	WA (%)	Lime	BD (g/cm^3)	Porosity (%)	WA (%)
10	1.61	37.4	39.3	10	1.07	-	31.4	10	1.09	-	-
30	1.80	39.1	38.9	30	0.97	-	45.1	30	1.18	-	-
50	1.86	36.2	37.2	50	1.02	-	32.7	50	1.21	-	-
70	1.92	34.4	31.6	70	1.03	-	26.8	70	1.24	-	-
90	1.95	42.3	28.5	90	1.23	-	24.2	90	1.29	-	-
100	2.04	37.4	26.1	100	1.24	-	26.8	100	1.22	-	-

The highest BD value (1.23 g/cm^3) for FA_{ISTPP}+MP mixture was found in the building panel specimen including 90% MP+10% FA_{ISTPP}. In the sample including 100% MP, the BD value is 1.24 g/cm^3. Since porosity tests cannot be performed on building panels produced from FA_{ISTPP}+MPA, they are not given in Table 5. The highest WA value (45.1%) in FA_{ISTPP} mixtures was determined in the sample having 30% MP+70% FA_{ISTPP} mixture. WA value was 26.8% in the building panel produced with 100% MP.

On the other hand, the highest BD value (1.29 g/cm^3) for FA_{ISTPP}+lime mixture was measured in 90% lime+10% FA_{ISTPP} and 80% lime+20% FA_{ISTPP} mixtures. The BD value of the building panel produced using 100% lime was 1.22 g/cm^3 (Table 5). Since porosity and WA tests cannot be performed on building panels produced from FA_{ISTPP}+lime, they are not given in Table 5.

4. Discussion and Conclusion

In the study, CS values of FA_{ISTPP}+lime building panels were decreased when the lime ratio was decreased. It was observed that water was split off from the mold surface when it is poured into molds by compression. This situation increases porous and cracked structure by decreasing FS and CS values.

It has been found that the FS and CS values of the building panels produced using 100% MP are higher than those produced using FA_{ISTPP}. The use of MP and lime with FA_{ISTPP} have reduced the values of FS and CS. This situation leads to opinion that the use of lime and MP was not suitable for producing high strength building panels. There has been an increase in FS and CS values when cement amount increased and FA_{ISTPP} decreased. For FA_{ISTPP}+lime building panels, CS values decreased when lime ratio decreased. On the other hand, as the FA_{ISTPP} amount increases in FA_{ISTPP}+MP building panels that there is a decrease in CS values. In the study of Kılınçarslan and Tuzlak (2018)[23], it was reported that the FA usage had a positive effect on the CS of foam concretes and stated that CS values increased in parallel with the increase in density. In their work, the scientists express that FA can be evaluated in foam concrete. In the present study, it was reported that BD values increase when the CS values of the building panels increase and FA_{ISTPP} can be used in the production of building panels. In this study, building panels produced by using FA_{ISTPP}+cement with less than 70% FA_{ISTPP} contributions and FA_{ISTPP}+MP with less than 50% FA_{ISTPP} contributions meet the related standards. In the study, reuse of FA_{ISTPP} to a building panel for the construction sector with the cleaner production principle was done. It is a study on industrial symbiosis. Since industrial solid waste (FA_{ISTPP}) is completely converted into product, it also supports the zero-waste philosophy. Having higher CS value is one of the most important features for building materials. The building panels produced in this study have high CS and low BD values, so they offer high strength in lightweight building material. If serially produced, it is envisaged that it will contribute to the solution of the problem of solid waste, the economy and human health.

5. Acknowledgment

This study was presented in the 1st International Conference on Environment, Technology and Management (ICETEM). We thank Prof. Dr. Ahmet Bilgil for his valuable comment and contribution. The authors would like to thank ÇİMSA, one of the leading companies of Turkish cement industry, which supports this study and Dr. Adnan Güven, former manager of the ÇİMSA Niğde Plant.
This work was partially supported bythe project "Development of eco-friendly composite materials based on geopolymer matrix and reinforced with waste fibers" funded by the European Commission, within the 7th Framework Programme for Research and Technology Development (FP7), Topic #02: Waste management, recycling and urban mining (Project No. ELAC2015/T02- 0721) under the ERANet-LAC: Latin America, Caribbean and European Union and supported by a grant of the Turkish Scientific and Technological Research Council of Turkey (TÜBİTAK). The Turkish project number is 116Y549.

References

[1] Ahmaruzzaman M 2010 *Prog. En. Combustion Sci.* **36** 327-363
[2] Sonebi M 2002 *Mater. Struct.* **35** 373-380
[3] Aitcin PC *et al* 1984 *The use of condensed silica fume in grouts(innovative cement grouting, SP-83)* American Concrete Institute, 1-18
[4] Ruggiero J G 1984 *Low slump compactive tail shield grouting in soft ground shield driven tunnels*(Innovative Cement Grouting, SP-83) American Concrete Institute, p. 103-114
[5] Weaver K D *et al* 1990 *Concr. Int.* **12** 45-47
[7] Tsivilis S *et al* 1999 *Cement. Concrete. Comp.* **21** 107-116
[8] Voglis N *et al* 2005 *Cement. Concrete. Comp.* **27**191-196
[9] Tsivilis, S*et al* 2002 *Cement. Concrete. Comp.* **24** 371-378
[10] Dhir R K 2007 *Construction. Mater. Struct.* **40** 459-473
[11] Rahhal, Vand Tolero R 2005 *Cement. Concrete. Res.* **35** 1285-91
[12] Yahia A *et al Cement. Concrete. Res.* **35** 532-539
[13] Poppe A M and Schutter G D 2005 *Cement. Concrete. Res.* **35** 2290-99
[14] Bentz D P 2006 *Cement. Concrete. Comp.* **28** 124-129
[15] What is Cement? History-Chemistry-Industries. In: Civil Engineering 2019. https://civiltoday.com/civil-engineering-materials/cement/81-cement-definition-and- full-details. Accessed 29 October 2019
[16] Moulding Plaster. In: ACG Materials 2019 https://www.acgmaterials.com/moulding-plaster/. Accessed 29 October 2019
[17] Baltar L M *et al* 2013 *Int. J. Miner. Process.* **125** 5-9
[18] Vimmrova A *et al* 2014 *Cement Concrete Comp.* **52** 91-96
[19] Plastering Applications. In: Graymonthttps://www.graymont.com/en/markets/building-construction/plastering. Accessed 29 October 2019
[20] Plastering Applications. In: Graymonthttps://www.graymont.com/en/markets/building-construction. Accessed 29 October 2019
[21] Kenny M and Oates T 2000 *Lime and Limestone* (Ullmann's Encyclopedia of Industrial Chemistry) ed Elvers B (Berlin)
[22] Boynton R S 1980 *Chemistry and Technology of Lime and Limestone* (New York: Wiley& Sons)
[23] Kılınçarslan Ş and Tuzlak F 2018 Investigation of strength and thermal conductivity properties of foam concrete with fly ash, *Uluslararası Sürdürülebilir Mühendislik ve Teknoloji Dergisi* **1** 1-5 (in Turkish)

IOP Conf. Series: Materials Science and Engineering **706** (2019) 012004 doi:10.1088/1757-899X/706/1/012004

Geopolymer's radiation absorption detected by Geiger-Muller counter

Elif Süzgeç[1,2], Neslihan Doğan-Sağlamtimur[1,2], Vakkas Bozkurt[3] and Sefa Ertürk[3]

[1]Department of Environmental Engineering, Nigde Ömer Halisdemir University, Turkey
[2] Nanotechnology Application and Research Center, Nigde Ömer Halisdemir University, Turkey
[3] Department of Physics, Nigde Ömer Halisdemir University, Turkey

E-mail: neslihandogansaglatimur@gmail.com, nds@ohu.edu.tr

Abstract. Fly ash (FA), which is an industrial solid waste released by the burning of coal in thermal power plants, is responsible for causing environmental problems. Radiation is also an environmental problem besides causing health problems for human being. Geopolymer is a specialized building material resulting from the reaction of a source material, which is rich in silica and alumina, with alkaline solution. It is ordinary Portland cement (OPC) free building material. It has been reported that geopolymer building material has good engineering properties with a reduced carbon footprint resulting from the total replacement of OPC as greener alternative. In this study, it is aimed to measure radiation absorption of geopolymers produced from the reaction of FA, which was supplied from İsken Sugözü Thermal Power Plant (Adana, Turkey) and rich in silica and alumina in total content of 83.74, with alkaline activators (sodium hydroxide and sodium silicate) at curing temperatures (70 °C and 100 °C). In the study, lattice measurement system consisting of nested lead plates to avoid unwanted radiation was designed to determine radiation absorption of geopolymers. By using this new system, geopolymer was found to be 5% better radiation absorption capacity than OPC building material when it is compared.

1. Introduction

In the case of waste disposal, the biggest problem facing industries is the safe and efficient disposal of by-products such as fly ash (FA) and blast furnace slag produced in large quantities during the combustion of waste, sediment and coal used. Most of this ash from thermal power plants is stored in appropriate areas and then dumped in landfills. Regular storage is not a highly preferred option because it not only causes a large financial burden for thermal power plants, but is also responsible for future environmental costs and regular storage arrangements. In addition to these problems, the increased burden of toxic metals in the storage area potentially increases the threat to groundwater contamination [1].

FA is driven out of the boiler with flue gases in coal-fired power plants and then generally captured from flue gases by some special equipment (e.g. electrostatic precipitator or other particle filtration equipment) before the flue gases reach the chimneys. It poses a major threat to the environment due to its adverse effects on biological diversity as well as its impact on groundwater [2].

IOP Conf. Series: Materials Science and Engineering **706** (2019) 012004 doi:10.1088/1757-899X/706/1/012004

One of the most important objectives in obtaining sustainable construction materials is to minimize the excessive use of virgin materials used to produce cement, coarse and fine aggregates and to minimize environmental damage. The use of industrial by-products such as FA, silica fume, ground granulated blast furnace slag and rice husk ash as a substitute for cement or as additional materials to cement has a constructive effect in minimizing greenhouse gas emissions. Millions of tons of industrial waste are produced each year, most of which is disposed of in the trash and very few are recycled. In recent years, there has been increased awareness of the amount and diversity of hazardous solid waste generation and its impact on human health. Increased concerns about the environmental consequences of waste disposal have led researchers to investigate the use of waste as potential construction materials.

In order to obtain an environmentally friendly and sensitive concrete, many studies are still going on for using waste concrete to produce green concrete. The most successful of the researches was the development of geopolymer concrete to eliminate the use of cement. FA-based geopolymer concrete was first introduced in 1979 by Davidovits to reduce the use of OPC in concrete [3].

Ordinary Portland cement (OPC) has been the most common cementitious material used to produce concrete, mortars, and cementitious composites in general. However, each ton of Portland cement releases about a ton of carbon dioxide into the atmosphere. The average world carbon intensity in cement production was assessed 0.81 kg CO_2/kg cement two decades ago. The greenhouse gas emission from the production of Portland cement is about 4 billion tons annually, which is about 7% of the global greenhouse gas emissions. The commitment towards limiting CO_2 emissions and stopping global warming has generated a burgeoning interest in finding alternative binders for OPC. Geopolymer concrete, obtained from different types of binder and having a completely different chemistry and microstructure than the one of OPC concrete, represents a promising alternative [4].

The radiation sources are generally occurred from cosmic and terrestrial radiation. Radiation sources are potentially harmful and they can be found in water, air, food and building materials [5]. The levels of radiation generally depend on the origin of the materials. Cement is an important construction material we use for buildings and it is known that some construction materials are naturally more radioactive. Therefore, radiation protection from natural radioactivity in building materials becomes important topic to be investigated [6-8]. Since the radioactivity is harmful to human health, the determination of natural radioactivity emitted from building materials is important for the protection of workers and public in a long term. In the present work, radiation absorption values of geopolymer building materials produced by using FA supplied from İsken Sugözü Thermal Power Plant (ISTPP) in Adana (Turkey), Rilem Cembureau standard sand alkali activators (Na_2SiO_3 and NaOH) were measured by using Geiger-Müller Counter Gamma-Scout system which is a device for the detection of all types of radiation such as alpha, beta and gamma rays.

2. Materials and Methods

In this study, geopolymer building materials were used as material.Some details about contents of these materials are given in Table 1. The measurement of the radiation absorption for the each geopolymer samples was performed using Geiger-Muller counter with Cs-137 radioactive source covered 4 mm thick lead shielding box and carried out in physics laboratory of Niğde Ömer Halisdemir University.

In order to measure the radiation absorption percentage of the geopolymer building materials, a lattice system was designed. The schematic design of experimental setup is shown in Figure 1.

IOP Conf. Series: Materials Science and Engineering **706** (2019) 012004 doi:10.1088/1757-899X/706/1/012004

Table 1. Properties of materials used in this study.

	Contents
ASTM C618 (2000) Class-F FA	SiO_2: 62.28, Al_2O_3: 21.46, Fe_2O_3: 7.01, CaO: 1.53, Na_2O: 0.26, MgO: 2.37, K_2O: 3.81, SO_3: 0.07, Others:1.21 in % BET: 2.26 m^2/g, Loss on ignition (LOI): 1.78%
Rilem Cembureau standard sand	Complied with TSE EN 196-1 (2009) specification
Alkali activator	NaOH/ Na_2SiO_3within the ratios of 1:1, 1:1.5, 1:2, 1:2.5 and 1:3 by weight, respectively.

The measurements were performed by reading the value on the counter that detects gamma rays emitted from the radioactive source. In the first step, the Geiger-Müller counter was placed outside the lead-shielded box, and then gamma-rays emitted from the radioactive source (Cs-137) placed inside the shielded box were detected (Figure 1.a) for 1 hour in the absence of geopolymer building materials. In the second step, a geopolymer building material was placed between the Geiger-Muller counter and radioactive source (Figure 1.b) inside the shielded box. Each measurement took for period of 1 hour, after the measurement the amount of radiation absorbed by each geopolymer specimen was determined (Table 2) proportionally.

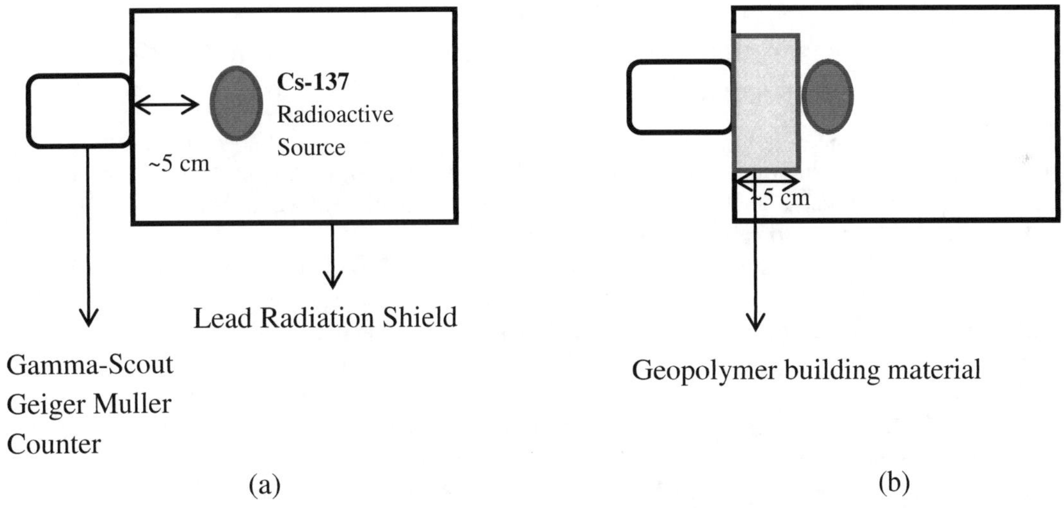

Figure 1. The schematic representation of experimental setup.

3. Results and Discussion

Mixing ratios of geopolymer building materials cured at 70 and 100 °C for 24 hours in an oven are given Table 2. Radiation counts detected by Geiger Muller apparatus and radiation absorption percentage calculated by using these counts of the geopolymer building materials are also given in this table. For reference measurement, average radiation count was determined as ~1083 for 1 hour in this setup when geopolymer was not exist.

Table 2. Mixing ratios, radiation count and radiation absorption percentages for different geopolymer building materials used in this study.

Sample Number	Curing Temperature (°C)	Mixing ratios of geopolymer building materials	Counting time (h)	Average count	Radiation Absorption Percentage (%)
Reference	-	-	1	1083	-
1	70	Na_2SiO_3 (%20)	1	1043	3.69
2		$NaOH/Na_2SiO_3$=1:1 (%20)		1050	3.04
3		$NaOH/Na_2SiO_3$=1:1 (%20)		1029	4.98
4		$NaOH/Na_2SiO_3$=1:1.5 (%20)		1026	5.20
5		$NaOH/Na_2SiO_3$=1:2 (%10)		1055	2.58
6		$NaOH/Na_2SiO_3$=1:2 (%20)		1050	3.04
7		$NaOH/Na_2SiO_3$=1:1.5 (%10)		1036	4.33
8		$NaOH/Na_2SiO_3$=1:1.5 (%20)		1053	2.77
9		$NaOH/Na_2SiO_3$=1:2.5 (%10)		1074	0.83
10		$NaOH/Na_2SiO_3$=1:3 (%20)		1070	1.20
11	100	$NaOH/Na_2SiO_3$=1:3 (%20)	1	1066	1.56
12		$NaOH/Na_2SiO_3$=1:1 (%10)		1033	4.61
13		$NaOH/Na_2SiO_3$=1:2 (%10)		1024	5.44
14		$NaOH/Na_2SiO_3$=1:1.5 (%10)		1036	4.33
15		$NaOH/Na_2SiO_3$=1:1.5 (%20)		1025	5.35

As seen in Table 2, radiation absorption values were found between 0.83 and 5.20% in geopolymer building materials produced at 70°C. The material produced with $NaOH/Na_2SiO_3$ in ratio of 1:1.5 (20%) had the highest radiation absorption capacity for this group.

For the geopolymer building materials produced at 100 °C, radiation absorption values were determined between 1.56 and 5.44%. The material produced with $NaOH/Na_2SiO_3$ in ratio of 1:2 (10%) had the highest radiation absorption performance as 5.44% for this group.

4. Conclusions

It is important to know whether the geopolymer building material is useful for radiation protection. Although the determination of radiation absorption in geopolymer building material is important, there are limited studies in the literature regarding this parameter. Binici et al. (2012) determined the physical, mechanical and radiation permeability and absorption properties of slag, FA, silica sand and pumice-based geopolymers [11]. It was measured that these geopolymers absorbed radiations in 5.9 and 17.3 keV energy. Only FA-based geopolymers absorbed 26 keV radiation effectively. Therefore, they could be used for radiation absorption material in the places using 26 keV radiation. On the other hand, Binici (2013) produced geopolymer mortars incorporating granulated blast furnace slag, FA, silica sand and pumice with NaOH in curing temperature of 100 and 150 °C for 20 hours [12]. In his study, geopolymer mortars were exposed to gamma and X-ray radiation. Water absorption, unit weight, flexural and compressive strength, adhesion properties of ultrasonic sound transmission and radiation absorption of the samples were investigated. Highest linear absorption coefficient was obtained from the sample incorporating blast furnace slag. It was reported that these geopolymer mortars may be used in the structures that are exposed to radiation.

IOP Conf. Series: Materials Science and Engineering **706** (2019) 012004 doi:10.1088/1757-899X/706/1/012004

In the present study, FA-based geopolymer building materials activated with NaOH and Na_2SiO_3 and cured at 70 and 100 °C were used as material. Radiation absorption of these materials were investigated. It was found that geopolymer building materials incorporating FA have radiation absorption capacity, especially due to their porous structure. The highest radiation absorption value (5.44%) was determined in the 13th sample produced in NaOH/Na_2SiO_3=2 mixture ratio and cured at 100 °C.As a result of this study, it was concluded that geopolymer building materials could be used in structures exposed to radiation. In this context, it is considered that the use of FA in geopolymer building materials will contribute to the protection of human health against radiation. Reusing FAs in order to produce geopolymer building materials have another advantage. Since FAs are converted into geopolymers and used in building material mixtures, they offer a great environmental advantage by solving waste storage problems. Furthermore, using geopolymer building materials is also expected to reduce CO_2 emission, contributing solution of the industrial solid waste problem and saving money for economy.

Acknowledgments
This study was presented in the 1st International Conference on Environment, Technology and Management (ICETEM). The geopolymer building materials of this study were produced in the project "Development of eco-friendly composite materials based on geopolymer matrix and reinforced with waste fibers" funded by the European Commission, within the 7th Framework Programme for Research and Technology Development (FP7), Topic #02: Waste management, recycling and urban mining (Project No. ELAC2015/T02-0721) under the ERA Net-LAC: Latin America, Caribbean and European Union and supported by a grant of the Turkish Scientific and Technological Research Council of Turkey (TÜBİTAK). The Turkish project number is 116Y549. This work was supported by this project.

This study includes partial sections of the project supported by TUBITAK 2209/B Programme, project grant no. 1139B411701722. The authors would like to thank Prof. Dr. Ahmet Bilgil, Associate Prof. Dr. Hatice Öznur Öz, Pervin Nas, Arife Gözde Akan, Betül Dikkaya, Türkan Vural and Turan Şevki Köker for their support during this study.

References
[1] Khale D and Chaudhary R 2007 *J. Mater. Sci.* **42** 729-746
[2] Ling Y 2018 *Proportion and performance evaluation of fly ash based geopolymer and its application in engineered composites* (PhD Thesis) Iowa State University, (Iowa, USA)https://lib.dr.iastate.edu/cgi/viewcontent.cgi?article=7405&context=etd. Accessed 8October 2019
[3] Islam A *et al* 2014 *Mater. Des.* **56** 833-841
[4] Bhutta A*et al* 2017 *Mater. Struct.* **50** 1-13
[5] ICRP 2015 Protection of the public against exposure ındoors due to radon and other natural sources of radiation: *Specific Safety Guide* No. SSG-32
[6] UNSCEAR United National Scientific Committee on the Effects of Atomic Radiation 2000 Sources and risks of ıonizing radiation, report to the general assembly with annexes (United Nations New York)
[7] Merle L and Enn R 2012 *Proceedings of the Proce. Estonian Acad. Sci.* **61** 107-112
[8] Zalewski M *et al* J 2001 *Pol. J.Environ. Stud.* **10** 183-188
[9] ASTM C618, 2000. Standard specification for coal fly ash and raw or calcined natural pozzolan for use as a mineral admixture in concrete
[10] TS EN 196-1 2009 Methods of testing cement-Part 1: Determination of strength
[11] Binici H *et al* 2012 *Yapı Tek. Elekt. Der.* **8** 12-25
[12] Binici H 2013 *Adv. Civ. Environ. Eng. (ACEE)* **1** 108-123

IOP Conf. Series: Materials Science and Engineering **706** (2019) 012005 doi:10.1088/1757-899X/706/1/012005

Production of geopolymer composites by using different alkaline solution/material ratio

Neslihan Doğan-Sağlamtimur[1,3], Hatice Öznur Öz[2,3], Ahmet Bilgil[2,3], Elif Süzgeç[1,3] and Türkan Vural[1,3]
[1]Department of Environmental Engineering, Nigde Ömer Halisdemir University, Turkey
[2]Department of Civil Engineering, Niğde Ömer Halisdemir University, 51240, Niğde, Turkey
[3] Nanotechnology Application and Research Center, Nigde Ömer Halisdemir University, Turkey

E-mail: oznuroz@gmail.com, oznurozell@gmail.com

Abstract. This study reports an experimental program to optimize mix design parameters of fly ash-based geopolymer engineering composites. Fly ash (FA) supplied from Çatalağzı Thermal Power Plant (Zonguldak, Turkey) and Rilem Cembureau Standard Sand were used together with a constant ratio of 0.50 for producing geopolymer. In the first step, two different alkaline solution/material ratio (FA+standard sand) (L/M) were determined as 0.20 and 0.40, respectively. After that, sodium silicate ((Na_2SiO_3) and 12M sodium hydroxide (NaOH) were utilized at a ratio of 1, 2, and 3, respectively. 6 mixtures obtained by using different design parameters were cured at 70 °C for 24 hours, respectively, and then they were stored at room temperature until testing age. While the physical properties of the samples were measured by the bulk density and water absorption tests at 28 days, mechanical properties were tested via the compressive and flexural strength tests at 90 days, respectively.

The results indicated that the highest compressive strength for geopolymer materials was found as 36.5 MPa in the mixture produced with the ratios of L/M=0.2 and Na_2SiO_3/NaOH=2 (by weight). However, less alkali reaction caused by using excessive activator solutions decreased the strength characteristics of geopolymer.

1. Introduction

The awareness of sustainable development in the world has led to the restructuring of characteristics properties for construction materials in the view of strength requirements as well as environmental impacts. It is known that approximately 4 billion tons of greenhouse gas emissions are emitted from the production of ordinary Portland cement (OPC); this corresponds to approximately 7% of global greenhouse gas emissions [1,2]. The use of industrial wastes such as fly ash (FA), blast furnace slag and silica fume as raw material instead of cement is advantageous both economically and environmentally to reduce the need for OPC [3,4]. In addition to eliminating the use of OPC, the growing concern of sustainable development for green concrete has led to the study of new building materials called as "geopolymers".

Geopolymer is produced with the composition of binding materials containing large amounts of silica and alumina and activated by the use of alkaline liquid [5,6]. Geopolymer production does not require the high temperature for the calcination, and also, the geopolymer can be produced from industrial by-products with minimal energy. Thus, compared to OPC production, the amount of greenhouse gas supplied to the environment decreased by 44-64% [7,8]. The desired characteristics of geopolymer is based on the composition of solid raw material depends on the molecular structure, the temperature and duration of the thermal curing or calcination process, and the reaction of the respective chemicals with the required molar mixing. FA is a suitable raw material for the geopolymerization technology obtained from thermal power plants by electrostatic or mechanical deposition of particles for geopolymer production since geopolymerization is linked to the alumina-silicate chain [9]. Komnitsas, (2011)[10] stated that the most common alkali activators used in the production of geopolymer materials are sodium hydroxide (NaOH), sodium silicate (Na_2SiO_3), and potassium hydroxide (KOH) or potassium silicate (K_2SiO_3). Komljenović et al. (2010) [11] found that the nature and concentration of the activator is the most dominant parameter in the alkali activation process. Additionally, the degree of hydration reactions can be increased by the alkaline concentration in the geopolymer mix and thus, pore volumes reduces by improving microstructural properties of the C-S-H product [12,13]. Also, Ling(2018) [13] determined that curing temperatures changed between 50-80 °C are widely acknowledged values used for successful geopolymer hydration. However, it was highlighted by the researchers that the concentration of activator has a greater effect on the mechanical strength than those of the curing temperature and time [14-16]. Moreover, Hardjito et al. (2017) [2] indicated that the increasing ratio of FA had a positive effect on the workability of FA based geopolymer concrete. Additionally, it was stated that the increasing curing period time even with low curing temperature enhanced compressive strength, tensile strength, and modulus of elasticity of geopolymers.

This study covers an experimental study to obtain the best design parameters for the production of alkali activated geopolymer material. For this purpose, 6 mixtures were designed with different liquid/material (FA and sand) and alkaline activator ratios (1, 2, and 3). At the end of the curing process, physical tests were determined via bulk density, water absorption tests while mechanical properties were obtained by compressive and flexural tests at the related testing age.

2. Materials and Methods

FA used in this study was obtained from Çatalağzı Thermal Power Plant in Zonguldak (FA). Table 1 presents the physical and chemical properties of FA.

Table 1. Chemical composition of FA.

Compounds	Amount (%)
CaO	1.77
SiO_2	42.12
Al_2O_3	20.27
Fe_2O_3	6.07
MgO	1.44
K_2O	3.81
Na_2O	0.26
Others	
Loss on ignition (LOI) (%)	1.52
Specific gravity	2.04
BET (m^2/g)	1.11

As seen Table 1, FA can be classified as class F type according to ASTM C618 (2000)[17]. A special sand called as Rilem Cembureau(specific gravity: 2.56, density: 1.35 kg/dm^3) (TSE EN 196-1) [18] was used as fine aggregate. The alkaline activator was comprised of Na_2SiO_3 solution (Na_2O=14.7%, SiO_2=29.4% andH_2O=55.9) and 12M NaOH solution. In the first step, two different alkaline solution/material ratio (FA+standard sand) (L/M) were determined as 0.20 and 0.40, respectively. And then, Na_2SiO_3 with NaOH in ratio of 1, 2, and 3 by weight were used for the design of geopolymer mortars, respectively. After producing the geopolymer mortar, the samples were cured at 70°C for 24 hrs. The details of mix design of the geopolymer mortar are given Table 2. The test batches were produced until obtaining the slump flow diameter in 16±2 cm.

In the preparing of geopolymer mixture, first of all, $Na_2SiO_3$and NaOH were mixed to get the determined ratio for the alkaline solution (Na_2SiO_3/NaOH) into the mixer. Then, FA and sand were added to the mixer and mixtured slowly for 5 minutes to achieve the homogeneity. The mixed mortar was poured into pre-prepared molds with the dimensions of 40x40x160 mm. Finally, the prepared geopolymer samples were placed in a laboratory oven for thermal curing at 70 °C for 24 hours. The mixtures given in Table 2 were coded based on the testing parameters such as the ratio of L/M, the ratio of Na_2SiO_3/NaOH. For example, if the L/M ratio and Na_2SiO_3/NaOH ratio were used as 0.2 and 1, respectively, this geopolymer mortars were denoted as 20FA-1.

Table 2. Geopolymer sample mixture properties.

FA (g)	Sand (g)	Alkaline Activator/Materials Ratio	Na_2SiO_3/NaOH Ratio (mL/mL)			Testing Temperatures (°C)
			1/1	1/2	1/3	
400	800	20%	120/120	80/160	60/180	70
		40%	240/240	160/320	120/360	

While the physical properties of the samples were measured by the bulk density and water absorption tests according to ASTM C642-13 (2013) [19] at 28 days, mechanical properties were tested via the compressive and flexural strength tests conducted by ASTM C 348-14 (2017) [20] and ASTM C 349-14 (2017) [21] at 90 days, respectively. The average of six measurements for each test is reported.

3. Results and Discussion

As seen in Figure 1, the bulk density values of geopolymer samples changed between 1842 and 1944 kg/m^3. The bulk density of hardened mortars was increased by the ratios of L/M. However, the highest bulk density values were obtained at the ratio of Na_2SiO_3/NaOH=2. This may be attributed to the change in homogeneity of the geopolymer matrix by the presence of fine sand [22].

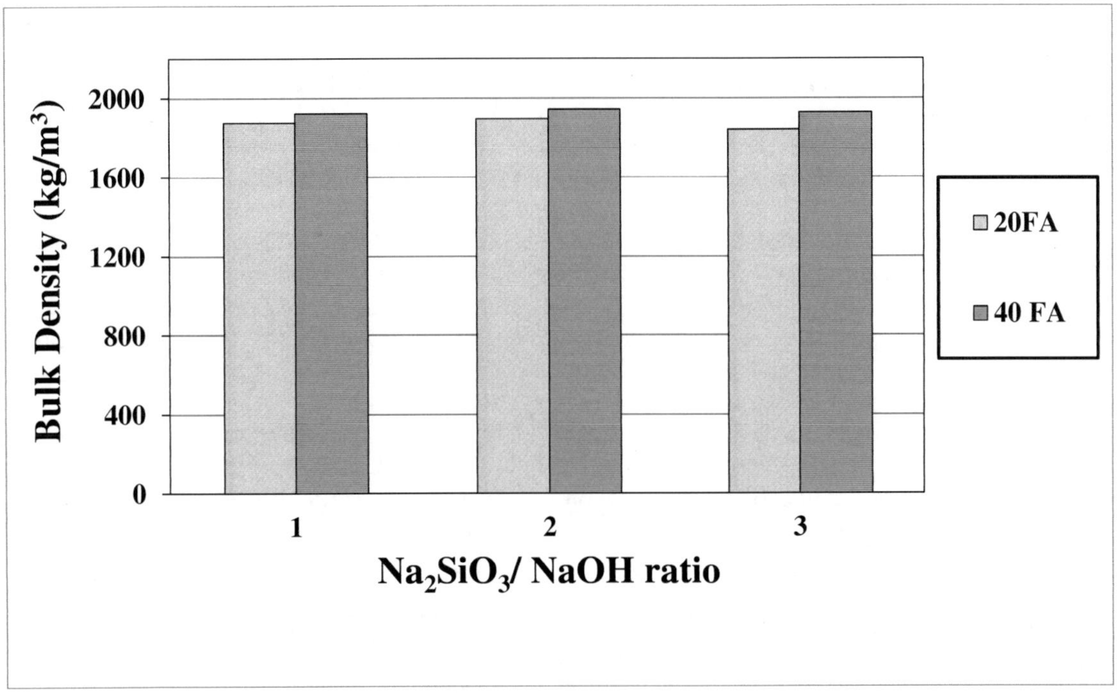

Figure 1. Bulk density of geopolymer mortars.

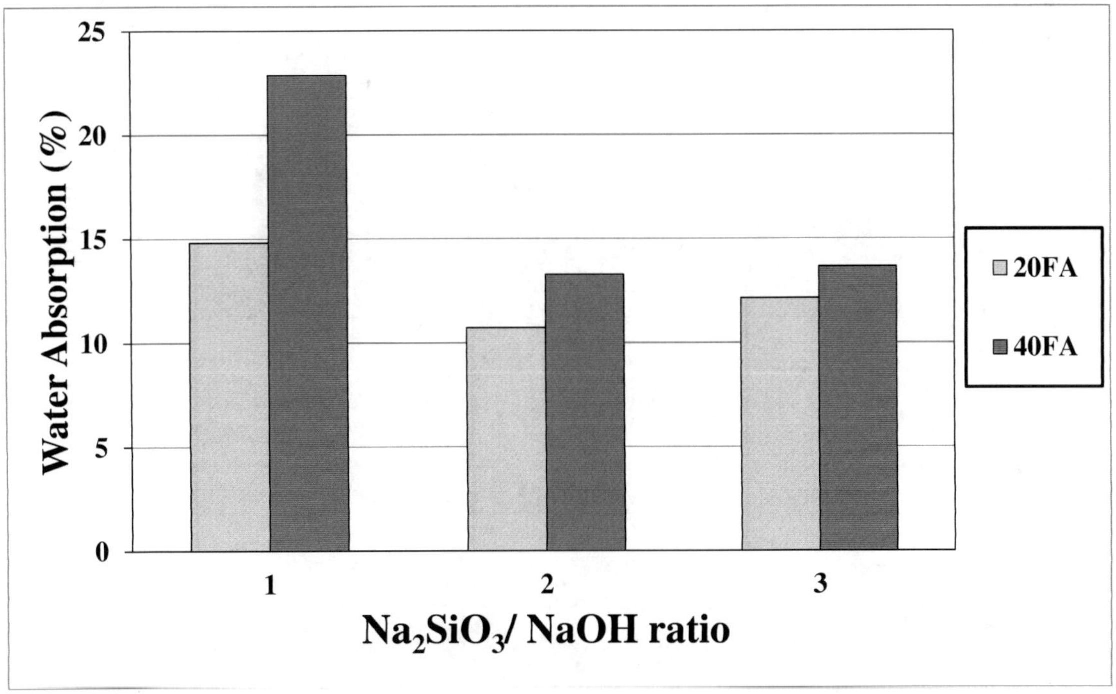

Figure 2. Water absorption capacity of geopolymer mixtures.

Figure 2 indicates that the water absorption values of geopolymer mortars measured at 28days with different design parameters. The lowest water absorption capacity of geopolymer samples produced at 70 °C curing temperature was found as 10.73% for 20FA-3 while 40FA-1 had the highest one

(22.87%). A seen in Figure 2, the increasing ratio of L/M resulted in an increment of water absorption capacity for the geopolymer mortars.

Compressive and flexural test results of mixtures are given in Figures 3 and 4 at 90 days, respectively. Despite the increasing of L/M ratio, excessive amount of fine sand gave a reverse effect for the characteristic properties of geopolymer. At 90 days, the lowest compressive and flexural strength values of geopolymer samples were 8.1 and 3.2 MPa for 40FA-1 while 20FA-2 had the highest values found as 36.5 and 8.7 for compressive and flexural test results, respectively.

Increasing of Na_2SiO_3/NaOH solution ratio up to 2 improved the reaction rate and lead to a less porous structures in geopolymer matrix. It may therefore result lower water absorption capacity, thus increases strength for the FA-based geopolymer [13, 23]. The reaction kinetics and the rate of crystallization were affected by the presence of soluble silica. Additionally, the higher percentages of soluble silica in geopolymer affected the dissolution of FA, thus promote the precipitation of larger molecular species, creating in a stronger Si rich gel formation which is responsible for strength development of the material with an enhanced density [13, 24]. But, as seen in Figures 3-4, there was an optimum limit for the Na_2SiO_3/NaOH solution ratio, such that exceeding this limit resulted the adverse conclusion for the strength characteristics of the geopolymer [25, 26].

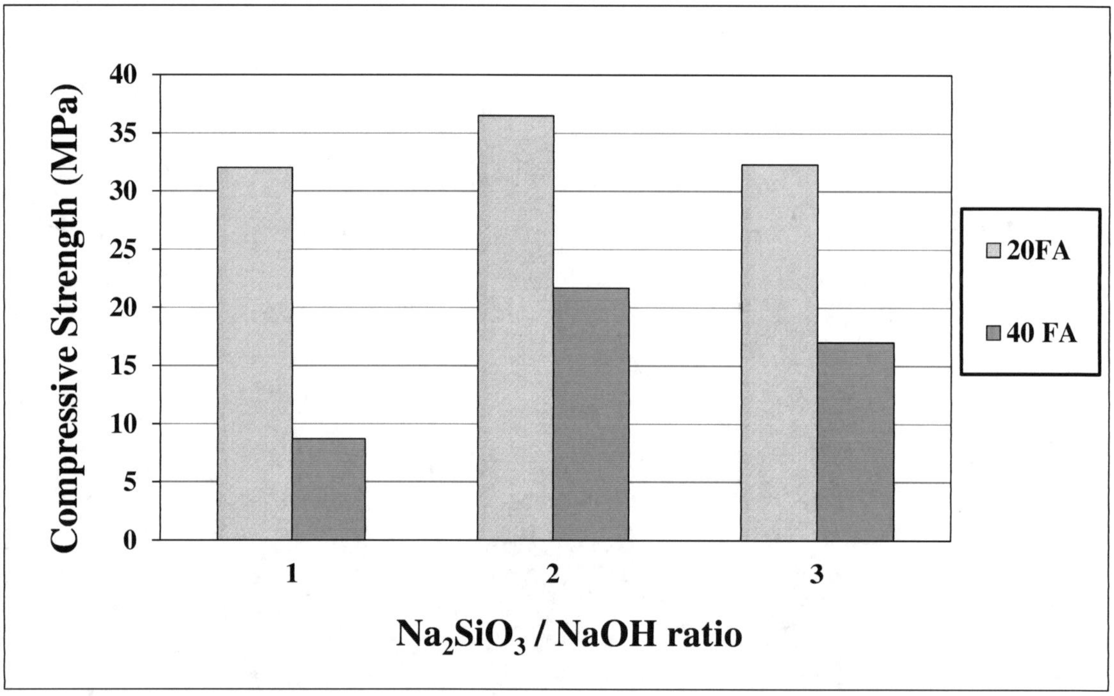

Figure 3. Compressive strength of geopolymer mixtures.

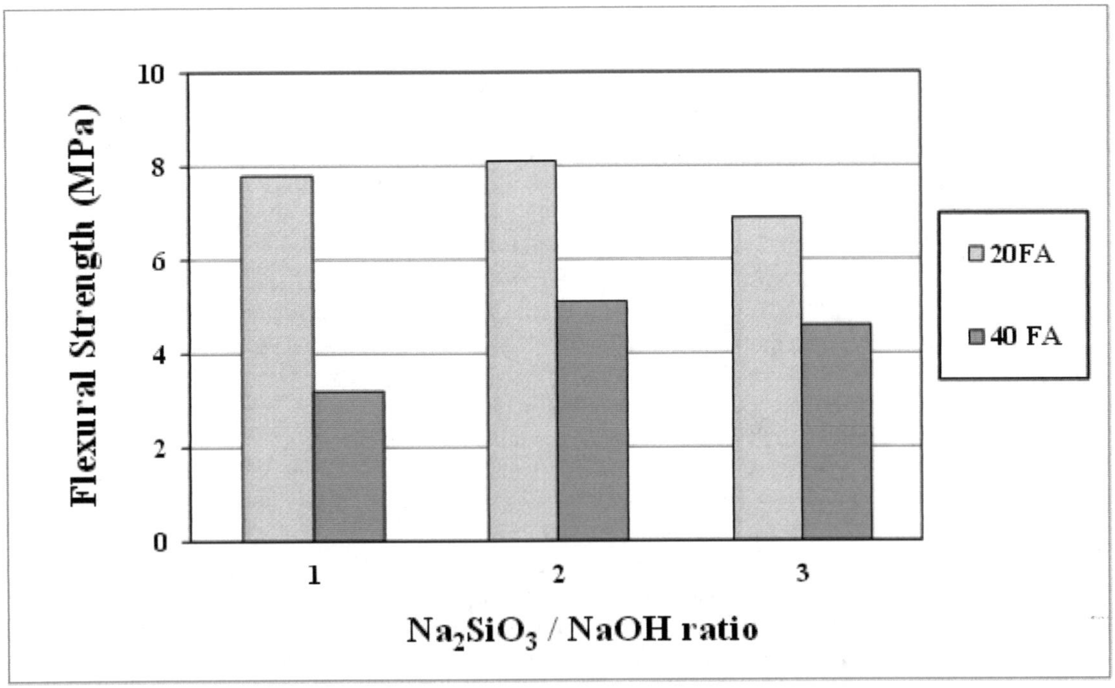

Figure 4. Flexural strength of geopolymer mixtures.

4. Conclusions

Based on the findings of the study, the following conclusions can be drawn:

- Producing geopolymer mortars with L/M ratio of 0.4 resulted in adverse effect for the water absorption, compressive and flexural strength of geopolymer when compared to the ones generated with lower one due to the excessive amount of fine sand.
- Increasing the ratio of Na_2SiO_3/NaOH gave an increment of strength characteristics up to an optimum point obtained as 2.
- Production of FA-based geopolymer not only provides an opportunity to use FA as a sustainable green material by alkaline activation but also avoids the disposal problems of FA. Additionally, this study contributes the sustainable development technology by decreasing CO_2 emissions into the atmosphere since it is a cement-free technology. Thus, geopolymers manufactured in this study with the desired strength characteristic can be used building bricks blocks, tiles, prefabricated materials as well as the infrastructure works.

Acknowledgments

This study was presented in the 1st International Conference on Environment, Technology and Management (ICETEM). It was supported by the project "Development of eco-friendly composite materials based on geopolymer matrix and reinforced with waste fibers" funded by the European Commission, within the 7th Framework Programme for Research and Technology Development (FP7), Topic #02: Waste management, recycling and urban mining (Project No. ELAC2015/T02- 0721) under the ERA Net-LAC: Latin America, Caribbean and European Union and supported by a grant of the Turkish Scientific and Technological Research Council of Turkey (TÜBİTAK). The Turkish project number is 116Y549. We thank Turan Şevki Köker, Arife Gözde Akan, Pervin Nas and Betül Dikkaya for valuable contribution in experimental parts of this study.

References

[1] Bhutta A *et al* 2017 *Mater. Struct.* **50** 1-13
[2] Hardjito D *et al* 2017 *ACI Mater. J.* **101** 467-472
[3] Islam A *et al* 2014 *Mater. Des.* **56** 833-841
[4] Khale D and Chaudhary R 2007 *J. Mater. Sci.* **42** 729-746
[5] Jumrat S *et al* 2011 *Int. J. Heat Mass Transf.* **38** 242-248
[6] Adam A A and Horianto X 2014 *Proce. Eng.* **95** 410-414
[7] McLellan B C *et al* 2011 *J. Clean. Prod.* **19** 1080-90
[8] Connie N *et al* 2018 *Constr. Build. Mater.* **186** 550-576
[9] Fekoğlu B 2018 Alternatifyapımalzemeleri, Available from:http://kisi.deu.edu.tr/burak.felekoglu/08.geopolimer.pdf. Accessed29 October 2019
[10] Komnitsas K A 2011 *Proce. Eng.* **21** 1023-32
[11] Komljenović *et al* 2010 *J. Hazard. Mater.* **181** 35-42
[12] Pacheco-Torgal F C *et al* 2007 *Constr. Build. Mater.* **22** 1315-22
[13] Ling Y 2018 Proportion and performance evaluation of fly ash based geopolymer and its application in engineered composites; Iowa State University; dissertation
[14] Hu M *et al* 2009 *Cem. Concr. Compos.* **31** 762-768
[15] Bakharev T 2005 *Cem. Concr. Res.* **35** 1224-32
[16] De Vargas *et al* 2011 *Cem. Concr. Compos.* **33** 653-660
[17] ASTM C618-00 2000 Standard specification for coal fly ash and raw or calcined natural pozzolan for use as a mineral admixture in concrete. *ASTM International* https://www.astm.org/DATABASE.CART/HISTORICAL/C618-00.htm.Accessed29October 2019
[18] TSE EN 196-1, 2009. Methods of testing cement-Part 1: Determination of strength
[19] ASTM C642-13 2013 Standard test method for density, absorption, and voids in hardened concrete
[20] ASTM C 348-14 2014 Standard test method for flexural strength of hydraulic-cement mortars
[21] ASTM C 349-14 2014 Standard test method for compressive strength of hydraulic-cement mortars (using portions of prisms broken in flexure)
[22] Kong D and Sanjayan J 2010 *Cem. Concr. Res.* **40** 334-339
[23] Fernandez-Jimenez A and Palomo A 2005 *Cem. Concr. Res.* **35** 1984-92
[24] Zuda L *et al* 2006 *Int. J. Thermo* .**27** 1250-1263
[25] Comrie D C and Kriven W M 2003 *Ceram. Trans.* **153** 211-225
[26] Nguyen K T *et al* 2017 *Constr. Build. Mater.* **130** 146-155

Fly ash as a raw material for geopolymerisation–mineralogical composition and morphology

K Korniejenko[1], M Łach[1], J Marczyk[1], C Ziejewska[1], NP Halyag[2] and G Mucsi[2]

[1] Institute of Materials Engineering, Faculty of Materials Engineering and Physics, Cracow University of Technology, Jana Pawła II 37, 31-864 Cracow, Poland.
[2]Institute of Raw Material Preparation and Environmental Processing, University of Miskolc, 3515 Miskolc, Egyetemváros, Hungary.

E-mail: kinga.korniejenko@mech.pk.edu.pl

Abstract. The article discusses the main features fly ashes (FAs) as a raw material for geopolymerisation taking into consideration mineralogical composition and morphology. It is continuation the previous research connected with chemical composition and physical properties. This article is focused on the examination of a FA from the CHP plant in Skawina (Poland) and assessment it as a main component for geopolymers production. The characteristic of the FA is presented, including the morphology and mineralogical structure. The morphology was examined using Scanning Electron Microscopy (SEM). The structure of the FA was monitored also by Fourier Transform Infrared Spectroscopy (FT-IR) and X-ray diffraction (XRD).

1. Introduction

A lot of mineral deposits in Europe are mostly depleted, reuse of waste and so called by-products is one of the main principles of sustainable development and circular economy [1, 2]. Additionally, growing environmental awareness and importance of development the sustainable construction materials with decreasing environmental impact for construction industry are main motivators to research works on possibility to use secondary raw material such as fly ash (FA) [3, 4].

Nowadays, the using of FA as so called secondary raw material becomes more and more popular, but still most FAs are deposited in open dumps without any pre-treatment. It involves a high risk of ecosystem pollution with danger to both human health and the environment [4, 5]. While the FA based geopolymer is an emerging alternative to conventional cement-based materials in civil engineering applications [1]. The particular motivation to implementation in the construction sector is planned eco-friendly effect. Currently, the building material production appears to be closer to traditional production models, characterized by high energy demand and high Green House Gases emissions, than other industries [2, 6]. These models should be turned into a green, circular and low-carbon economy [2, 7]. New solution such as using FAs and other wastes streams, including waste materials form construction and demolish industry and mine tailings as raw materials for geopolymerisation process help in contribution this sector in circular economy [1, 6, 8].

For the effective using waste materials such as FAs in geopolymerisation their quality is important [9, 10] as well as their safety for human health and the environment [11, 12]. The previous studies show that the chemical composition as well as the physical properties of the FA from the CHP plant in

Skawina (Poland) are proper for using this material for geopolymer synthesis. The chemical composition of this FA is fulfil the requirements for a class F according to ASTM678 standards (ASTM-C618-2). This FA contains less than 5% of unburned material, less than 10% of iron compounds and a low amount calcium content [13, 14]. The amount of the reactive silica is about 36% - it is under optimal value: 40 to 50%. The limits for content of heavy metals and toxic elements for this FA is much below standards [12]. It has also good physical parameters -the content of particle under the size 45 μm is about 88%.

But not only the physical and chemical properties influenced on the behaviour of fresh and hardened geopolymer also the crystallographic and morphology of the FA play important role [15, 16]. This paper analyses the mineralogical and morphology of the FA from the coal power plant 'Skawina' as a raw material for geopolymer production.

2. Material and methods

2.1. Material

The FA was delivered from the coal power plant 'Skawina' (located in: Skawina, Lesser Poland, Poland). The base material is bituminous coal. The power plant generates electricity and heat in its combined production. This FA is obtained by electrostatic precipitation of fine particles from the exhaust gases from coal-fired furnaces. According to the Polish law [17, 18], the FA is classified as a combustion by product with the no. 10 01 02 (Fly ash from coal).

2.2. Research methods

The mineralogy composition was determined using powder X-ray diffraction Debye-Scherrer Hull method. Diffractograms of samples were recorded with the use of a Rigaku SmartLab X-ray diffractometer. The following parameters were applied: CuKα radiation, a graphite reflection monochromator, tube voltage 45 kV, tube current 200, step scan mode: step size = 0.05°2θ, count time per 1 step = 1 s. The identification of phases based on data contained in the ICDD Catalogue (International Centre for Diffraction Data 2014) and XRAYAN software. The structure of the FA was determined by Fourier Transformed Infrared Spectroscopy (FT-IR). The samples induced by infra range electromagnetic waves was detected by JASCO FT-IR 4200 type Fourier Transformed Infrared Spectrometer in reflection mode, diamond ATR was used.

The morphology of the samples was examined by means of the scanning electron microscopy (SEM) type JEOL JSM 820 with EDS. The FA had been properly prepared before i.e. dried to the constant mass, put on the coal base to carry the electric charge of the sample away, and covered with thin layer of gold with the JEOL JEE-4X vacuum sputter. The observations were made at various magnifications (between 20 and 1000 times).

3. Results and discussion

3.1. Mineralogical characterization of FA (FT-IR, XRD)

The mineralogical composition is partly connected with the chemical composition and physical properties of the FA, especially content of reactive silica. The provided tests allow to observe non-crystalline phases in the FA. In the literature, this kind of phases are differently called by different authors: amorphous phase, vitreous phase or glassy phase [19-21]. However, there is some differences according chemical point of view between these terminologies [22] in the article there will be use appellation - amorphous phase.

The results of X-ray diffraction analysis of the FA from coal power plant 'Skawina' are presented in figure 1.

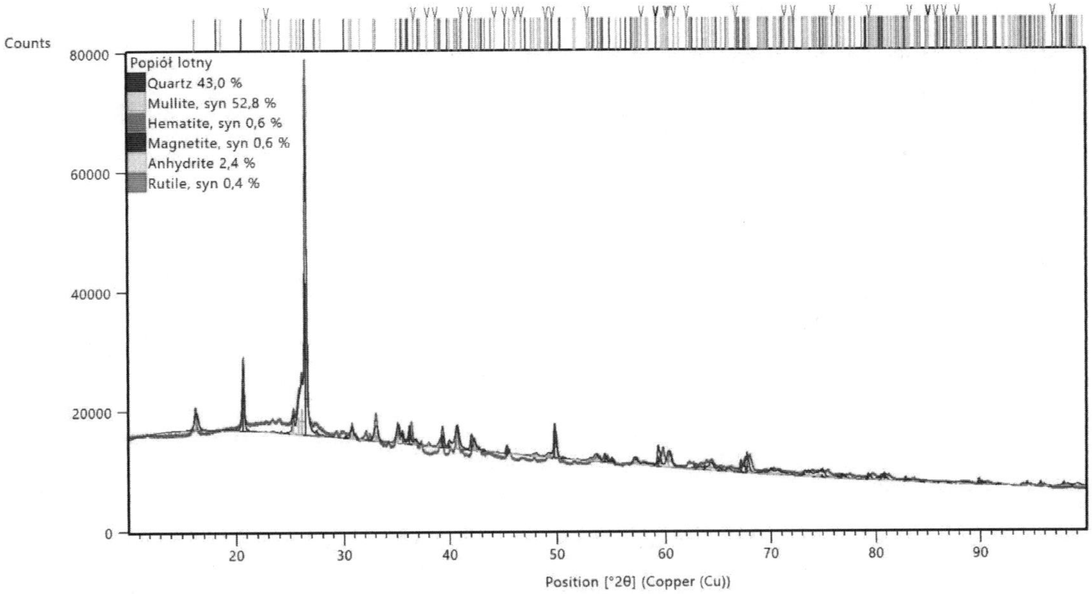

Figure 1. XRD pattern of the FA from the CHP plant in Skawina (Poland).

The identified phases were fitted to the diffractogram of the examined FA - table 1.

Table 1. The marked phases.

Name	Chemical formula	[%]	Ref. Code
Quartz	SiO_2	43,0	01-070-3755
Mullite	$Al_6Si_2O_{13}$	52,8	00-015-0776
Hematite	Fe_2O_3	0,6	01-084-9870
Magnetite	Fe_3O_4	0,6	04-006-6550
Anhydrite	$CaSO_4$	2,4	01-085-6133
Rutile	TiO_2	0,4	00-034-0180

The results show the presence of crystalline phases typical for FAs such as: quartz (related with silicon dioxide), mullite and anhydrite. The curves indicated also on presence amorphous phase (indicated by diffuse halo / curves has a broad hump). The presented composition is typical for FAs. Usually, the most frequent crystalline phases are: quartz, feldspar, mullite, anhydrite and hematite [19]. In case of application in geopolymer manufacturing the most important is large amount of amorphous phase. During the reaction, the reactive components undergo geopolymerisation and create binders whereas the un-reactive components behave like aggregate in traditional concretes [20].

To confirm of presence the aluminosilicates in the FA, especially amorphous phases the FT-IR measurements has been done. The FT-IR was made for the FA from the CHP plant in Skawina (Poland) - bituminous coal and compared with two coals form Hungarian: form brown coal and lignite [23, 24]. The results are shown in figure 2.

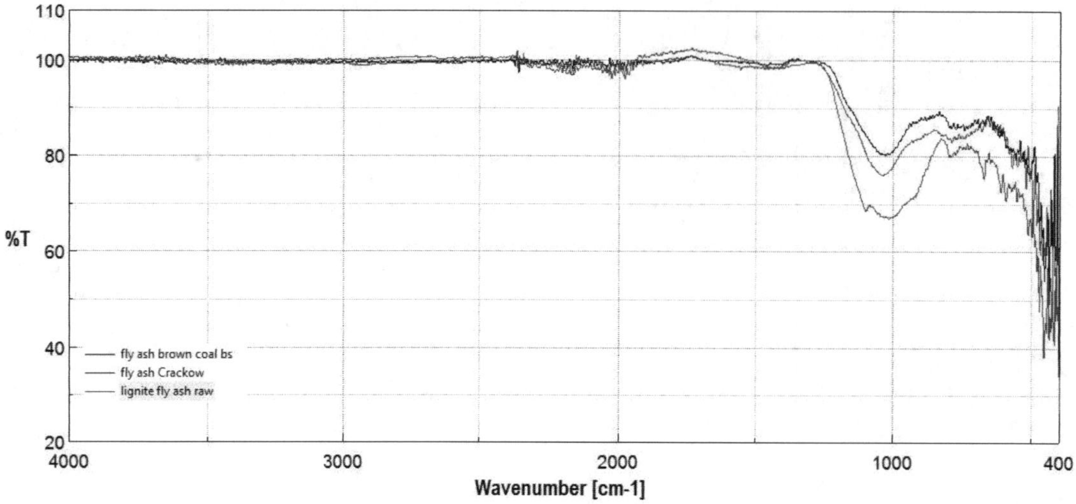

Figure 2. The FT-IR for the FA from the CHP plant in Skawina (Poland) with comparison to Hungarian FAs.

FT-IR clearly shows the presence of aluminium silicates and/or aluminosilicates in all FAs. The results show the domination of bands associated with the silica dominated in the all spectra. These are: the most intense band at about 1100 cm-1which is connected with asymmetric stretching vibrations and also Si-O(Si) bound at about 460 cm-1 connected with bending vibrations O-Si-O presented in silicate tetrahedra [25, 26]. The doublet about 800 cm-1 is associated with the symmetric stretching vibrations of Si-O-Si bridges. It is probably connected with low temperature quartz appearance that is lower in the FA form CHP plant 'Skawina' then for other FAs.A high full width at half maximum (FWHM) indicates on significant amount of amorphous phase which structure is silicate and/or aluminosilicate [25, 26].

Moreover, the band at around 1000 cm-1 appear in the analysed spectra. It may be connected with asymmetric stretching vibration of Si-O (Al) presented in aluminosilicates structures. Substitution of aluminium atoms in tetrahedral positions in these structures results in a change the position of the band: from the position about 1100 cm-1, in the case of 'pure' silica structure, to lower and lower wavenumbers with increasing Al/Si ratio [25, 27]. The Al/Si ratio increases with the decreasing values of wavenumbers due to the substitution of aluminium atoms for Si at the tetrahedral position [25, 28, 29].

The bound at about 915 cm-1 confirms the presence of aluminium in the octahedral position, confirming - the existence of mullite [27, 29]. Additionally, it is worth to notice presence the bounds associated with calcite (at about 1450 and 875 cm-1) and other bounds appearing at around 3400 cm-1 and 1625 cm-1 are attributed to the stretching vibration and bending vibration of OH in H_2O molecules. The second one indicates a small amount of molecular water [25].

Taking into consideration geopolimerisation process one of the most important parameters in amount of amorphous phase. The amount of amorphous phase is connected with content of reactive material in the FA. The 'reactivity' does not only refer to the amorphous phase, but to the FA as a whole [19, 30]. However, the amorphous SiO_2 content, amorphous Al_2O_3 and the median particle size of the FA can be used as reactivity index of FA used for geopolymers [21]. That means the increasing amount of amorphous content is reflected in a high proportion of so-called reactive silica [25, 26]. Usually, the FAs form coal has between 46 and 50% reactive silica [31] and amount of amorphous phase could be between 30 and 80% [32].

The FA reactivity and solubility in an alkaline solution also depends on the composition and content of the amorphous phase [19, 33]. The higher amount of amorphous phase is connected with lower Si/Al ratio and higher amount of soluble aluminium [19,34]. For the geopolymerisation process

FAs with different reactivity are expected to produce diverse reaction products, resulting in different compressive strengths [35, 36]. Especially, low reactivity of FA results in slow binding and early strength development. It caused delayed setting and lack of development full 3D network in material structure - the dissolution of the FA does not complete before the final hardened structure is formed [20, 28, 37]. In case of FAs that have high amount of reactive content, when there are mixed with the alkaline dissolution, the amorphous component is quickly dissolved. In such a situation, main reaction product is an amorphous or semi-crystalline material. The higher the amount of amorphous phase in the FA, the faster the activation process and the higher the degree of reaction in geopolymers is. It significantly influence on the mechanical strength of the final products [38, 39].

From the mineralogical point of view, together with increasing the content of amorphous phase, the capacity of activation for the FA is higher [38]. The content of amorphous phase is related with thermodynamic and kinetics of the activation process and affects the amount of aluminosilicates which is dissolved in the alkaline medium as well as to its rate of dissolution [38]. Finally, the reactivity of the material is connected with coal burning. The high temperature caused that the process the calcination has a place and the calcined materials such as the FA are mainly amorphous and has higher reactivity during geopolymerisation than uncalcined materials [37]. It is caused that calcination activates materials from crystalline to amorphous structure and extra energy stored in them [36, 37]. The amount of amorphous phase is also connected with the quick cooling of the FA from sinterised state. It makes the FA to be basically amorphous with small amounts of crystalline components [38]. That mean the reactivity depends strongly on the burning parameters for the coal in CHP plant.

3.2. Morphology

Moreover, is worth to notice that for the reactivity of the FA could influence not only the amount of the amorphous phase, but also the physical properties of the FA such as: the particles size - the best up of 45 μm (an important effect on the reaction rate of FA at early stages) [5, 40] and the surface area of the particles (considerable role in determining the kinetics of different processes) [5, 40]. Due to the fineness of FA particles the reactivity level increases [19, 41], with the surface area of the particles the issue is a little bit more complicate, because the process depends not only on measured value but also on the particles morphology [5].

The morphology of the particles of fly ash was typical of such by-products of coal combustion and suitable for the process of alkali-activation [16]. Observations of the morphology of the FA from the CHP plant in Skawina (Poland) particles have shown that they have regular shapes, mostly spherical (figure 3). From the point of view of the geopolymerisation process, the presence of spherical particles is beneficial and improves the rheological properties of the mixture, in particular, improves its workability and reduces the need for liquid substances [40], which in turn has a positive effect on seasoning and ensures the receipt of geopolymer materials with good mechanical properties. Additionally, EDS analyses confirmed the high content of key elements for the geopolymerisation process – silica and aluminium and the low calcium content (figure 4andfigure 5).

EDS analyse was performed for the selected approximation in point (figure 4) and in a selected area (figure 5). They were carried out on selected particles of the FA visible on the SEM image (figure 3). The main element is oxygen due to the composition of the FA, which is composed of oxides of various elements. Analyses show a large amount of silicon and aluminium in the ash and low calcium content. In addition, in point 1 (figure 4), the analysis showed a significant amount of iron, suggesting that it is visible in figure 4 on the "raid" sphere, iron oxides are present in the FA. Analyse in the area 3 (figure 5) show a small amount of this element. It is coherent with results of investigation for mineralogical composition - XRD. The FA also contains admixtures of other elements, i.e. magnesium in small amounts.

IOP Conf. Series: Materials Science and Engineering **706** (2019) 012006 doi:10.1088/1757-899X/706/1/012006

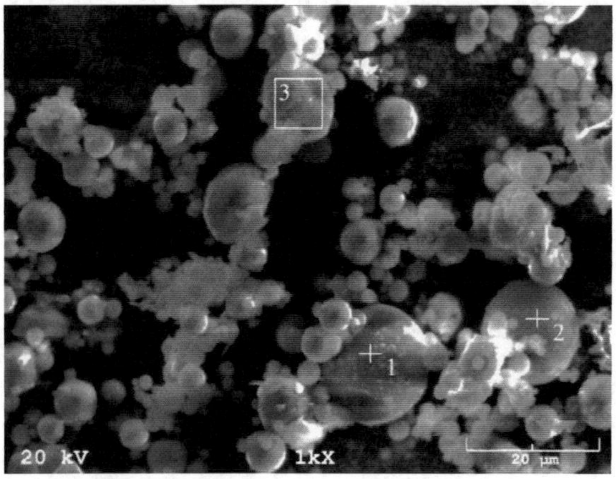

Figure 3. SEM image – the FA from the CHP plant in Skawina (Poland) [16].

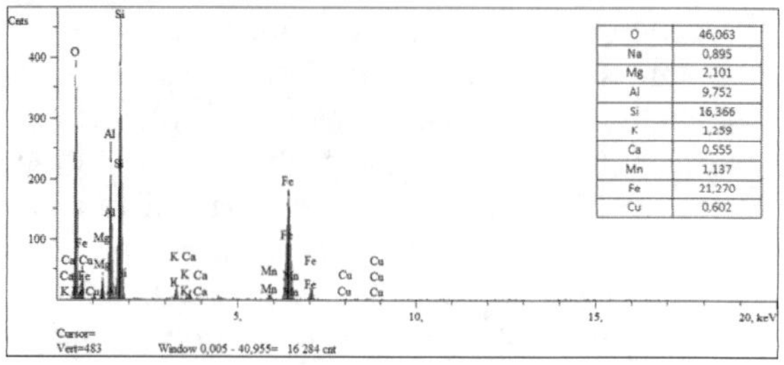

Figure 4. EDS analysis – chemical composition for the point no 1 on the figure 3 [16].

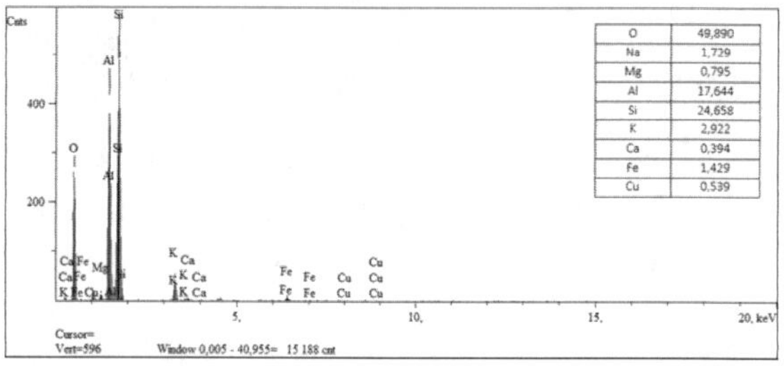

Figure 5. EDS analysis – chemical composition for the area no 3 on the figure 3 [16].

The properties of the FA based geopolymer could be affected by many parameters. A lot of them is significantly related to the primary materials and their characteristics. The particular parameters are often related each other, exemplary size and distribution of particles, the amorphous phase in the content and the reactivity of both silicon and aluminium are very often related with burning process [5, 28]. Other parameters such as: curing temperature and pressure, duration of curing, type of curing (conventional heating or microwave heating), type and concentration of alkalies or alkaline liquid-to-raw material ratio, are connected with geopolymerisation process. All of them must be properly combined to receive required properties of the final product [43, 44]. Usually, the manufacturer do not have influence on the quality of the raw materials, but it could implement some modifications for increasing their usefulness for geopolymerisation process trough pre-treatment or selection the process' parameters. Because of that it is very important to have a knowledge about the raw materials for proper design the geopolymerisation reaction.

4. Conclusions

The reactivity of the FA (defined as the ability of the FA to react with alkali solution and form geopolymers) is the combined effect of particle size and content as well as amorphous phase composition. The analyses confirm the usefulness of FA from the CHP plant in Skawina (Poland) for geopolymerisation.

The morphology examined by Scanning Electron Microscopy (SEM) shows that it is suitable for the process of geopolymerisation. The FA contains a lot of spherical particles and should have good workability and reasonable requirements on liquid substances. At the same time the mineralogical structure was investigated by Fourier Transform Infrared Spectroscopy (FTIR) and X-ray diffraction (XRD). The research shows the composition typical form FAs with large amount of amorphous phase, what is required in geopolymers bounding creation. The article do not presented the mechanical and physical properties of the FA that have been studied in other manuscript.

It can be concluded that the use of the FA from the CHP plant in Skawina (Poland) as a raw material in the production of geopolymers is highly recommended. It has required properties for this process and utilization this by-products, without down-cycling, for a production of construction materials. The quality of this secondary raw materials is comparable or in some case even better than compared to primary mining products such as metakaolin. Because of that it could have s significant contribute to circular economy and environmental protection development as well as possible energy savings and CO_2 emission reductions.

Acknowledgements
1. The research was made partly by the support of Tempus Public Foundation in the framework Bilateral State Scholarships.
2. This work has been financed by the by Polish National Agency for Academic Exchange under the academic exchange programme as part of bilateral cooperation with Hungary.
3. This work has been supported by the ERA Net-LAC 2nd Joint Call (http://www.eranet-lac.eu) funded by the National Centre for Research and Development in Poland, within the framework of the grant: *Development of eco-friendly composite materials based on geopolymer matrix and reinforced with waste fibers*.

References
[1] Paiva H, Yliniemi J, Illikainen M, Rocha F and Ferreira V M 2019 *Sustainability (Switzerland)* **11** (4) 995
[2] Orsini F and Marrone P 2019 *J. Clean. Prod.* **241** 118380
[3] Belviso C 2018 *Prog. Energ. Combust.* **65** 109
[4] Mikuła J and Korniejenko K 2015 *Innovative, Cost-Effective and Eco-friendly Fibre-based Materials for the Construction Industry* (Kraków: Wydawnictwo Politechniki Krakowskiej)
[5] Bouaissi A, Li L Y, Moga L M, Sandu I G, Al Bakri Abdullah M M and Sandu A V 2018 *Rev. Chim. (Bucharest)* **69** (7) 1661
[6] Favier A, De Wolf C, Scrivener K and Habert G 2018 *A sustainable future for the European Cement and Concrete Industry* (Zurich: ETH Zurich)
[7] Mikuła J 2014 *Rozwiązania proekologiczne w zakresie produkcji. Nowoczesne materiały kompozytowe przyjazne środowisku* (Kraków: Wydawnictwo Politechniki Krakowskiej)
[8] Silva G, Castañeda D, Kim S, Castañeda A, Bertolotti B, Ortega-San-Martin L, Nakamatsu J and Aguilara R 2019 *Constr. Build. Mater.* **215** 633
[9] Bajare D, Vitola L, Dembovska L and Bumanis G 2020 *Frontiers in Materials* **6** 22
[10] Bumanis G, Vitola L, Bajare D, Dembovska L and Pundiene I 2017 *Ceram. Int.* **43** 5471
[11] Sas Z, Sha W, Soutsos M, Doherty R, Bondar D, Gijbels K and Schroeyers W 2019 *Sci. Total Environ.* **659** 1496
[12] Mierzwiński D, Łach M, Mikuła J and Goły M 2018 *E3S Web of Conferences* **44** 00110
[13] Łach M, Korniejenko K and Mikuła J 2016 *Procedia Engineer.* **151** 410
[14] Łach M, Korniejenko K, Hebdowska-Krupa M and Mikuła J 2018 *MATEC Web of Conferences* **163** 06005
[15] Diaz E I, Allouche E N and Eklund S 2009 *Fuel* **89** 992
[16] Korniejenko K 2019 *Wpływ dodatku włókien krótkich na właściwości mechaniczne kompozytów na osnowie geopolimerowej* (Kraków: Biblioteka Cyfrowa Politechniki Krakowskiej)
[17] Act of 14 December 2012 on waste, J. of Laws of 2013, item 21, with later changes
[18] Regulation of the Minister of the Environment of December 9, 2014 regarding the waste catalog, J. of Laws 2014 item 1923
[19] Nikolić V,Komljenović M, Baščarević Z, Marjanović N, Miladinović Z and Petrović R 2015 *Constr. Build. Mater.* **94** 361
[20] Kumar S, Mucsi G, Kristály F and Pekker P 2017 *Adv. Powder Technol.* **28** 805
[21] Hadi M N S, Al-Azzawi M and Yu M 2018 *Constr. Build. Mater.* **175** 41
[22] Queiroz CA and Šesták J 2010 *Phys. Chem. Glasses: Eur. J. Glass Sci. Technol. B* **51** (3) 165
[23] Mucsi G, Molnár Z and Kumar S 2014 *Acta Phys. Pol.* A **126** (4) 994
[24] Mucsi G, Szabó R, Ambrus M and Kovács B 2018 *Analecta Technica Szegedinensia* **12** (1) 30
[25] Mozgawa W, Król M Dyczek J and Deja J 2014 *Spectrochim. Acta* A **132** 889

[26] Liu Y, Zeng F, Sun B, Jia P and Graham I T 2019 *Minerals* **9** 358
[27] Hahn A, Vogel H, Andó S, Garzanti E, Kuhn G, Lantzsch H, Schüürman J, Vogt C and Zabel M 2018*Sediment. Geol.* **375** 27
[28] Mukherjee S and Srivastava S K 2006*Energy Fuels* **20** 1089
[29] Fernández-Jiménez A and Palomo A 2005*Mesoporous Mater.* **86** 207
[30] Chen-Tan N W, van Riessen A, Ly C V and Southam D C 2009 *J. Am. Ceram. Soc.* **92** 881
[31] Kurdowski W 2010 *Chemia cementu i betonu* (Warszawa: Stowarzyszenie Producentów Cementu Kraków, PWN)
[32] Girgiczny Z 2006 *Rola popiołów lotnych wapniowych i krzemionkowych w kształtowaniu właściwości współczesnych spoiw budowlanych i tworzyw cementowych* (Kraków: Wydawnictwo Politechniki Krakowskiej)
[33] Brouwers H J H and van Eijk R J 2002 *J. Mater. Sci.* **37** 2129
[34] Brindle J H and McCarthy M J 2006 *Energy Fuels* **20** 2580
[35] Oh J E, Jun Y and Jeong Y 2014 *Cement Concrete Comp.* **50** 16
[36] Mucsi G, Kumar S, Csőke B, Kumar R, Molnár Z, Rácz A, Mádai F and Debreczeni Á 2015 *Int. J. Miner. Process.* **143** 50
[37] Komintsas K and Zaharaki D 2007*Miner. Eng.* **20** 1261
[38] Fernández-Jiménez A and Palomo A 2003 *Fuel* **82** 2259
[39] Jang J G and Lee H K 2016 *Constr. Build. Mater.* **102** 260
[40] Jayant D 2013 *Mineral Admixtures in Cement and Concrete* (Boca Raton, Florida: Taylor & Francis Group)
[41] Komljenovic M, Baarevic Z and Bradic V 2010 *J. Hazard. Mater.* **181** (1-3) 35
[42] Chindaprasirt P 2011 *Fuel* **90** 2118
[43] Lelusz M 2012 *Budownictwo i Inżynieria Środowiska* **3** (4) 179
[44] Giergiczny E and Giergiczny Z 2010 *Cement Wapno Beton* **15/77** (3) 157

IOP Conf. Series: Materials Science and Engineering **706** (2019) 012007 doi:10.1088/1757-899X/706/1/012007

Mechanical properties of composites based on geopolymers reinforced with sizal

K Korniejenko[1], M Łach[1] and J Mikuła[1]

[1] Institute of Materials Engineering, Faculty of Materials Engineering and Physics, Cracow University of Technology, Jana Pawła II 37, 31-864 Cracow, Poland.

E-mail: kinga.korniejenko@mech.pk.edu.pl

Abstract. The aim of the study was to evaluate the usefulness of composites based on alkali activated materials reinforced with natural fiber - sisal, as an ecological material for selected applications in the building industry. Microscopic analysis of the created structure and examination of compression and bending strength were carried out. Strength tests show a slight increase in compression and bending strength compared to a geopolymer matrix without fibers.

1. Introduction

Natural fibers used as a filler have many advantages over other fibers, including plastic ones. They are not only an environmentally friendly alternative, but also have lower density, lower cost and no toxicity [1, 2]. Additionally, these fibers are easy to use in processing and in many cases it is possible to use them in the technology park already owned by the companies [1]. At the same time, these fibers usually provide sufficient mechanical properties for most desirable applications in the field of construction [2,3]. In particular, due to the above-mentioned advantages, they are more and more widely used as a component of composites, inter alia, based on cements and geopolymeric materials [4,5]. Currently, research is conducted on natural materials such as cotton [1,6], flax [7,8], pineapple leaf [9], sorghum [10], coconut [2], raffia [2], wool [11], waste from the wood industry [12,13] and other organic fibers [2,14]. One of the possible additives to geopolymeric matrices are sisal fibers [2,9]. Sisal fibers come from the plant Agave Sisalana, which was originally found in Mexico [15,16]. Currently, the main producers of sisal fibers are Brazil, Mexico, Kenya, Tanzania, Colombia, Madagascar, China, Cuba, Haiti, Nicaragua, Angola, Indonesia, Mozambique, South Africa, and Thailand [15]. An unquestionable advantage of this plant is the possibility of growing all year round in hot climates and dry regions, which are often unsuitable for other crops. It is a low-maintenance plant and can be grown on most types of soils, it is also relatively resistant to diseases. The plants can be harvested from 2 years after planting, and their life span can be up to 12 years, producing 180 to 240 leaves depending on the conditions of cultivation. The yield per hectare is about 1 tone, although yields in East Africa can reach as much as 4 tons per hectare [16,17]. The fibers are made from plant leaves, contain about 90% moisture, are stiff and the fleshy flesh is very hard [15]. The fibers are deposited lengthwise in the leaves, especially near their surface. The first step in obtaining them is to separate them from the rest of the leaves as soon as they have been cut. This aims to avoid the risk of damage during the cleaning process. The fibers are removed either by mechanical peeling or by hand by separating the fibers from the rest of the leaf.

Sisal fibers have traditionally been used for cordage products. Nowadays, they are mainly used in the automotive industry, as a filler for plastic composites and in the paper industry. These fibers are also used in the production of: massage articles, orthopedic equipment, fabrics, as well as furniture or wallpapers [17,18].

It is also worth noting that sisal is a renewable source par excellence. During its entire life cycle, sisal absorbs more carbon dioxide than it produces. During processing, the plant generates mainly organic waste and leaf residues, which can be used to generate bioenergy, produce animal feed or fertilizer [19]. It is 100% biodegradable. In addition, sisal, as a plant with an extensive root system, is used to reduce soil erosion, as well as a barrier to protect farmland and forests from animals.

2. Material and methods

2.1. Material

Fly ash from the Skawina CHP plant was used as the material to create the geopolymer matrix. The oxide composition of the ash used is shown in Table 1.

Table 1. Oxide composition of fly ash from the Skawina CHP plant [4].

	LOI	SiO_2	Al_2O_3	Fe_2O_3	CaO	MgO	Na_2O	K_2O	SO_3	TiO_2	P_2O_5	BaO
[%]	2,44	55,89	23,49	5,92	2,72	2,61	0,59	3,55	0,16	1,09	0,82	0,20

Sisal fibers were added to the composite as a filler. The basic chemical properties of these fibers are known from the literature [20]. The basic component of the fiber is cellulose and it constitutes about 45.3% by volume of the fibers, the other components are: hemicellulose about 20.4%, lignin about 15.1% and other components about 15.5%. The basic physical and mechanical properties are presented in Table 2.

Table 2. Physical and mechanical properties of sisal fibers [16,21].

Property	Value
Density	1,5 g/m³
Young's modulus	9,0–38 GPa
Tensile strength	363–700 MPa
Elongation	2,0–7,0%

The added fibers had a diameter of about 0.5 mm and were mechanically shredded into pieces of 3 mm in length. The fibers were added in the volume of about 1% by weight in relation to fly ash used as the base for the geopolymer matrix. The amount of added fiber was determined on the basis of previous research on artificial and natural fibers [4, 22].
Crushed sisal fibers in 3% by weight were added to fly ash. Sodium hydroxide (NaOH) solution with a molar concentration of 8M containing sodium water glass was introduced into the mixture prepared in this way. 50 x 50 x 200 mm samples were prepared for flexural strength tests.
To produce geopolymers flakes of technical sodium hydroxide were used and an aqueous solution of sodium silicate (R-145) whose molar ratio was 2.5 and density was about 1.45 g/cm^3. The tap water was used instead of the distilled one. The alkaline solution was prepared by means of pouring the aqueous solution of sodium silicate over the solid sodium hydroxide. The solution was thoroughly mixed for 15 minutes in a low-speed mixer and allowed to equilibrate until a constant concentration and temperature. Next, the obtained paste was poured into cubic molds 50 x 50 x 50 mm. The solidification was conducted on the vibratory table. Then the molds were heated for 24 hours at 75°, cooled to the ambient temperature, taken out of the molds and stored for 28 days.

IOP Conf. Series: Materials Science and Engineering **706** (2019) 012007 doi:10.1088/1757-899X/706/1/012007

2.2. Research methods

The morphology of the samples was analyzed on the material remaining after the strength tests. The JEOL JSM-820 scanning microscope was used for the tests. Samples were previously properly prepared.Small amounts of materials dried to the constant mass, put on the coal tape to carry the electric charge of the sample away, and covered with thin layer of gold with the JEOL JEE-4X vacuum evaporator. Observations were carried out at various magnifications (50 - 2000 x).

Compressive strength tests according EN 12390-3, using Matest 3000 kN, were conducted on cubic samples of 50 × 50 × 50 mm conditioned at room temperature for 28 or 90 days. Five repetitions were carried out for each type of sample.

The bending strength tests were conducted according to EN 12390-5 on Instron Type 4465 Universal Testing Machine on 50 x 50 x 200 mm cubic specimens stored at ambient temperature for 28 days.

3. Results and discussion

Microscopic investigations allowed to observe the microstructure of the fiber and the obtained composite. The fiber structure (Fig. 1) and the filler-matrix connections (Fig. 2-3) are presented in the images.

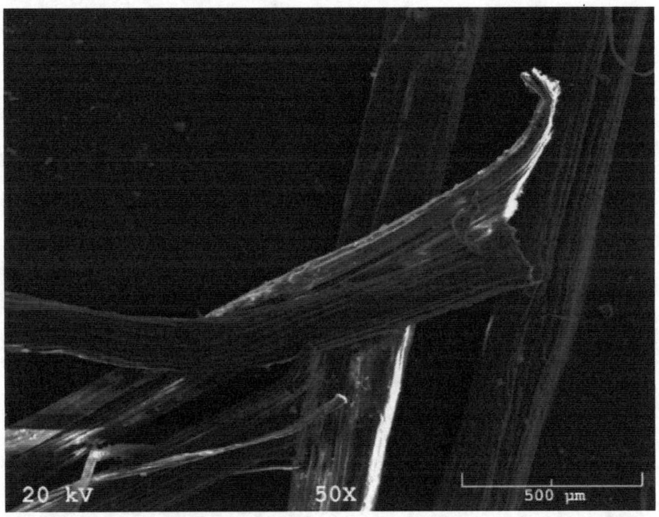

Figure 1. SEM image of used sisal fibers.

Figures 2 and 3 show the fibers in the geopolymer matrix. The fibers are evenly distributed in the whole volume of the composite, they are dispersed - distributed without a defined directional order. Such a way of fibers arrangement allows to obtain mechanical properties of the product similar in all planes.

IOP Conf. Series: Materials Science and Engineering **706** (2019) 012007 doi:10.1088/1757-899X/706/1/012007

Figure 2. SEM image obtained for a geopolymeric composite reinforced with 1% sisal fibers showing the distribution of fibers in the composite.

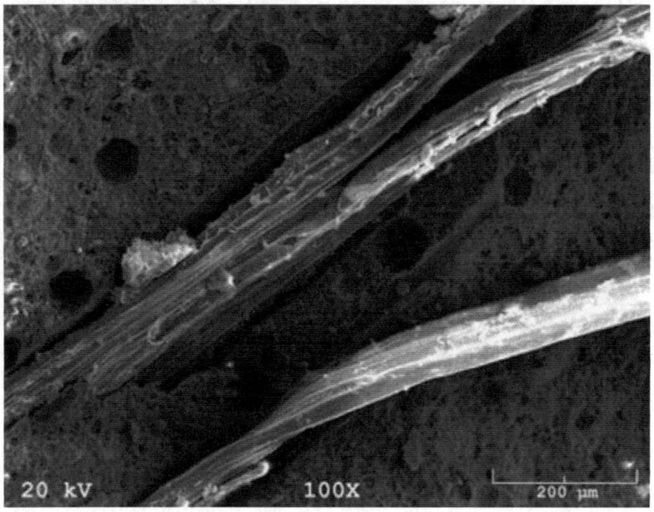

Figure 3. SEM image obtained for a composite on geopolymer matrix reinforced with 1% sisal fibers showing the cohesion of the fibers with the matrix .

SEM analysis allowed to observe the uniform distribution of fibers in the whole composite and good fiber - matrix cohesion (Fig. 2 – 3).

3.1. Results of strength tests

The results of compression and bending strength tests were carried out on previously prepared composite samples. Geopolymer without fiber addition was used as a material for comparison. The tables show average values for particular materials.

The results of compressive strength tests for geopolymer with added fiber are slightly better compared to the material of geopolymer matrix itself (Table 3).

Table 3. Results of the compressive strength tests [2].

Sample	Compressive strength [MPa]	Standard deviation
Geopolymer without fibers	24,78	1,89
Geopolimer with 1% sizal fibers	25,16	3,43

The bending strength was also improved with the addition of fibers. The results of the bending strength tests for geopolymer with added fiber are slightly better than for the material of the geopolymer matrix itself (Table 4).

Table 4. Results of flexural strength tests [2].

Sample	Flexural strength [MPa]	Standard deviation
Geopolymer without fibers	5,55	0,72
Geopolimer with 1% sizal fibers	5,90	0,14

Strength properties tests indicate a slight increase in the strength properties of composites with 1% addition of sisal fiber.

Conclusions

The paper presents the results of preliminary research works concerning the influence of the addition of sisal fibers in 1% weight volume on the geopolymer properties. The presented results confirm the potential of new material for selected building applications. Microscopic observations indicate good quality of connections between geopolymer matrix and sisal fiber. Strength tests show a slight increase in compression and bending strength compared to the geopolymer matrix.
There is now a demand on the market for new environmentally friendly composite materials, in particular those based on renewable raw materials. These materials should provide an environmentally friendly alternative to traditional construction materials used in the widely perceived construction industry. Modern composites allow to reduce the emission of substances harmful to the environment and at the same time save natural resources. Geopolymeric composites with natural fiber fillers are part of the sustainable development policy, which is currently a guideline in the policy of many countries in Europe and worldwide.

Acknowledgements

1. This work has been supported by the ERA Net-LAC 2nd Joint Call (http://www.eranet-lac.eu) funded by the National Centre for Research and Development in Poland, within the framework of the grant: *Development of eco-friendly composite materials based on geopolymer matrix and reinforced with waste fibers.*

References

[1] Alomayri T, Shaikh F U A and Low I M 2013 *Journal of Material Science* **48** 6746–6752
[2] Korniejenko K, Frączek E, Pytlak E and Adamski M 2016 *Procedia Engineering* **151** 388–393
[3] Kidalova L, Stevulova N, Terpakova E and Sicakova A 2012 *Journal of Clean Production* **34** 116
[4] Mikuła J, Korniejenko K. (eds.), *Innovative, Cost-Effective and Eco-friendly Fibre-based Materials for the Construction Industry*, Wydawnictwo Politechniki Krakowskiej, Kraków, 2015
[5] Alomayri T, Shaikh F U A Low I M 2013 *Composites: Part B* **50** 1–6

[6] Alomayri T, Shaikh F U A and Low I M 2014 *Composites: Part B* **60** 36–42
[7] Alzeer M and MacKenzie K 2013 *Applied Clay Science* **75–76** 148–152
[8] Assaedi H, Alomayri T, Shaikh F U A and Low I 2015 *Journal of Advanced Ceramics* **4** (4) 272–281
[9] Correia E A, Torres S M, Alexandre M E O, Gomes K C, Barbosa N P and Barros S D E 2013 *Materials Science Forum* **758** 139-145
[10] Chen R, Ahmari S and Zhang L 2014 *Journal of Material Science* **49** 2548–58
[11] Alzeer M and MacKenzie K J D 2012 *Journal of Material Sciences* **47** 6958–65
[12] Yan L, Kasal B and Huang L 2016 *Composites Part B: Engineering* **92** 94-132
[13] Yliniemi J, Nugteren H, Illikainen M, Tiainen M, Weststrate R and Niinimäki J 2016 *International Journal of Mineral Proc.* **149** 42-49
[14] Pacheco-Torgal F and Jalali S 2011 *Construction and Building Materials* **25** (2) 575
[15] Santos J D G, Vieira I J C, Braz-Filho R and Branco A 2015 *International Journal of Molecular Sciences* **16** 8761-71
[16] Silva F A, Chawla N and Toledo Filho R D 2008 *Composite Science and Technology* **68** 3438 43
[17] Dlamini T S, Verschoor A J and Tshabalala P 2014 *Journal of Development and Agricultural Economics* **6** (9) 386-393
[18] Gutierrez A, Rodriguez I M and del Rio J C 2008 *Industrial Crops and Production* **28** 81-97
[19] Cantalino A, Torres E A and Silva M S 2015 *Journal of Agricultural Science* **7** 64-74
[20] Fangueiro R and Rana S (eds.), Natural Fibres: Advances in Science and Technology Towards Industrial Applications, Springer, 2016
[21] Rana S S, Gupta M K and Srivastava R K 2015 *Materials Today: Proceedings* **2** 2909 – 17
[22] Korniejenko K, Mikuła J and Łach M., 2015, Characterization of mechanical properties of short glass fiber-reinforced geopolymer composites, ICCST/10 : 10th International Conference on Composite Science and Technology, 2-4 September 2015, Lisboa, Portugal

IOP Conf. Series: Materials Science and Engineering **706** (2019) 012008 doi:10.1088/1757-899X/706/1/012008

Electrochemical impedance spectroscopy analysis of iodine/potassium iodide electrolyte influence on spin-coated platinum counter-electrode on FTO glass

J Walter[1], D Mierzwiński[1], K Marszałek[2] and J Sanetra[3]

[1] Cracow University of Technology, Warszawska 24, 31-155 Krakow, Poland
Faculty of Materials Engineering and Physics
[2] AGH University of Science and Technology, Mickiewicza 30, 30-059 Krakow
Faculty of Computer Science, Electronics and Telecommunications
[3] retired scientist, earlier: Cracow University of Technology, Warszawska 24, 31-155 Krakow, Poland

Email: janusz.walter@mech.pk.edu.pl

Abstract. Very low concentrated precursor solution has been carried out for producing catalytic platinum layers by a spin-coating method. Direct deposition of platinum nanoparticles from low concentrated precursor solution on a conductive substrate results in many nanoparticles remaining in solution. They are not adhesive enough to be able to load them onto another submerged element. The spin-coating method has been applied to use them for catalytic layer preparation. Electrochemical impedance spectroscopy measurements were carried out using a symmetrical setup with the test electrolyte showing, in general, the better properties if more layers are applied. The circuit model for the symmetric setup, containing two identical Pt layers with electrolyte between them, has been proposed. It was found deteriorating layer properties expressed by charge transfer resistance increasing while the platinum layers were influenced by the electrolyte due to a decrease in the amount of platinum on the electrode surfaces. Layers of platinum retain their properties for some hours in contact with electrolyte without measurable decreasing of properties. The spin-coated platinum layer obtained by described method can be used in laboratory tests of Dye Sensitised Solar Cells different components filled with iodide-triiodide electrolyte.

1. Introduction

Dye sensitised solar cells (DSSC) may consist of various kinds of counter-electrodes (CE) which are crucial for the cell operation [1-3]. The efficiency of the processes, which require catalysts, depends on the active surface of the catalyst. Platinum is frequently used to improve the charge transfer from the fluorine doped tin oxide glass (FTO) to the electrolyte in DSSC cells and can be also used in various application as a catalyst [4]. A reduction of the costs associated with the use of a platinum, has led to the search for platinum with an extremely high surface area. The application of platinum nanoparticles provides very good catalytic properties and a relatively low cost of material. The counter electrode for a DSSC application is usually prepared by spreading hydrogen hexa-chloroplatinate isopropanol solution on FTO glass substrate [5].

One of the frequently used methods for obtaining platinum nanoparticles is a bottom-up method known as the polyol process [6]. Platinum nanoparticles with a size of a few nanometers, characterised by very good catalytic properties, can be formed during this process. They can be used as a catalyst in dye sensitised solar cells [7, 8, 9]. The polyol process is based on the chemical reaction between the polyols and precursors, which are the source of nanocrystalline platinum [10, 11]. The most widely used precursors are the platinum chloride $PtCl_4$ or potassium hexa-chloroplatinate K_2PtCl_6. The concentration of the nanoparticles in solution, after the process, depends on the amount of precursor used. Using high concentrations of the precursor results in a high material cost. That is suitable for the preparation of pastes for use in a screen-printing technique and economically justified because pastes may consist all formed nanoparticles. This method of obtaining platinum layers can be applied in mass production.

Fabrication of platinum catalytic layers on small scale can be realised by direct deposition of platinum on the substrate, during their formation process, by immersing the substrate in the solution while the polyol process is in progress. This process requires much less of the precursor compare to its concentration required for mentioned screen-printing technique and is very convenient for a laboratory tests. However only a small amount of the formed platinum nanoparticles from low concentration of precursor solution are deposited on the conductive substrate during the process, providing its electrocatalytic properties. Many nanoparticles remain in solution after direct deposition. Remaining nanoparticles are not adhesive enough to be able to load them onto another submerged element. These particles can be used for the preparation of an electrocatalytic layer applying spin-coating method. The proposed method of obtaining catalytic layers based on remaining nanoparticles reported in this study is a very convenient procedure for producing catalytic layers and useful for laboratory tests. Once obtained nanoparticles can be stored in a vessel and used for preparing an electrocatalytic layer when it is needed. The usage of spin-coating of remaining in the solution nanoparticles provides good repeatable catalytic properties for an experimental DSSC cell. Electrochemical impedance spectroscopy (EIS) is frequently used for the evaluation of electrode properties, fuel cell's characterisation or for checking the properties of different capacitive objects [12, 13] as well as for the assessment of dye sensitised solar cells and counter-electrode properties [14, 15]. EIS was used in our work to estimate the electrocatalytic properties of obtained platinum layers. The influence of the platinum layer time in contact with the electrolyte was tested to find long time stability of layer properties.

Mechanisms of the electrocatalytic properties of platinum have been reported in the literature. One of these results showed platinum instability in iodide/triiodide redox couple [15]. However, over the past several years, the iodide-triiodide electrolyte has been the most frequently used as contained the universal redox couple, because of its satisfactory kinetic properties. Other researchers have shown some disadvantages of Pt, like high cost, low abundance and poor corrosion resistance against an iodine based electrolyte. It was reported that the Pt reaction with iodine ions, which forms $PtCI_4$ or H_2PtCI_6, causes a degradation of the electrocatalytic activity of platinum [16, 17]. Other research has shown vapour deposited Pt on FTO does not seem to be stable in an electrolyte containing dissolved LiI and I, while solid Pt wire was found to be stable, and did not dissolve, after exposure to the electrolyte [18]. That may be a result of the nano-size of the vapour deposited Pt. Other work showed a drop of current density, for triiodide reduction, of up to 40% caused by the storage of Pt obtained from thermal decomposition of H_2PtCl_6 in electrolyte solution [19].

2. Material and Methods

2.1. Material

Nanoparticles were prepared by a polyol process using 0.001 M solution of platinum chloride $PtCl_4$ (Merck), and polyethylene glycol PEG600 (Fluka). The process was carried out for obtaining platinum layer on immersed before process FTO glass at the temperature of 380K for a duration of 60 min. During the heating process platinum nanoparticles were formed in solution and deposited on conductive side of FTO glass. Remaining in solution nanoparticles was further used for layer preparation using spin-coating technique.

2.2. Methods

X-ray Powder Diffraction (XRD) measurements of the produced nanoparticles were conducted to check the crystal lattice structure and its parameters. A spin-coating technique, at 3000 rpm for a duration of 20 s, was used, sequentially, to produce one, two and three layers of platinum nanoparticles on conductive side of FTO glass (TEC 15), drying each layer at 390 K before the next loading. Remains of the PEG600 in produced layers were finally removed in the end by heating at a temperature of 720 K for 30 minutes. Distribution of the nanoparticles on the surface was analysed using energy-dispersive spectrometer (EDS) IXRF Systems Model 500 Digital Processing, and the light transmittance of each layer was checked before the main EIS measurements. The layers were used to prepare symmetric cells for EIS two electrode measurements. The cells consisted of two identical Pt layers, formed on the FTO glass, which were facing each other. The distance between FTO-Pt electrodes, each of which were 1 cm^2 in surface area, was about 25 μm, with this gap being filled with electrolyte (figure 1). Similar cells, used by other researchers, have been reported in the literature [20, 21, 22]. The benefit of these cells is that only a small volume of electrolyte is required to conduct measurements. The cell volume was enclosed in Surlyn 1702 heat-sealing foil (Dupont) which separated the electrolyte from any atmospheric influence. The EIS investigations were conducted using a basic electrolyte based on iodine and potassium iodide. The final composition of the test electrolyte contained 0.5 M potassium iodide and 0.05 M of iodine in a solution of polyethylene glycol (PEG 600). Measurements were taken consistently up to after 40 days (961 h) to check the influence of the electrolyte on the electrocatalytic properties of platinum. The cells were then disassembled to measure the transmittance of light through the platinum layers influenced by the electrolyte to check the surface coverage by the platinum nanoparticles.

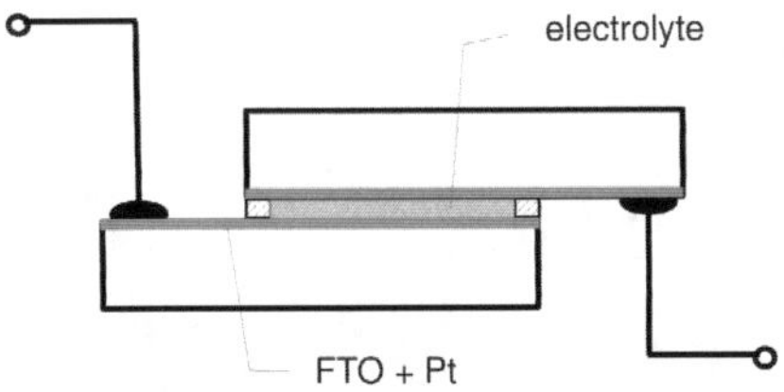

Figure 1. Electrochemical cell used for the measurement of impedance spectrum.

The crystal lattice structure of the platinum nanoparticles, produced by the polyol process, was examined using a Bruker diffractometer using ICDD (International Centre for Diffraction Data 2016). The crystal lattice parameter of the platinum was calculated using the Bruker Diffrac software from the half-peak width of the diffraction pattern. The distribution of nanoparticles has been checked on JEOL JSM 820 Scanning Electron Microscope on the FTO glass and analysed by the image analysis method on a surface of approx. 20 μm^2 using ImageJ software. Mapping time, at a magnification 15000 x, was set to 1.5 h for all samples to verify platinum loading.

EIS examinations were carried out using Zahner IM6e device to characterise the differences between cells (I, II or III layers of Pt) and check the catalytic properties of the platinum layers and stability of them over time. Measurements were conducted in the frequency range of 1Hz to 1 MHz, applying a 5 mV AC voltage amplitude. The electrical circuit model for obtained fitting the curves has been proposed using the SIM module Zahner IM6e electrochemical workstation. The same, or similar, models have been proposed in other research [20, 21, 22].

3. Results and discussion

The results of X-ray examination are shown in figure 2. Diffraction tests showed platinum-specific peaks. The exact value of crystal lattice parameter was equal to 0.3923 nm, which was consistent with the database. The result of this calculation showed that the crystalline size was about 6 nm. Analysis of the distribution of platinum on the FTO glass surface was carried out using EDS to compare the loading of the platinum after the spin-coating cycles. An almost homogeneous coverage of the FTO substrate was found with the tested layers having different amounts of nanoparticles on each sample, corresponding to the number of applied spin-coating cycles (I, II or III cycles). The mapping results are shown in figure 3. The result of EIS investigations of the symmetrical cell setup (figure 1) of the FTO plate, with one layer of platinum using the spin-coating technique is shown in figure 4. Cells built using two and three layers of platinum are shown, respectively, in figure 5 and figure 6.

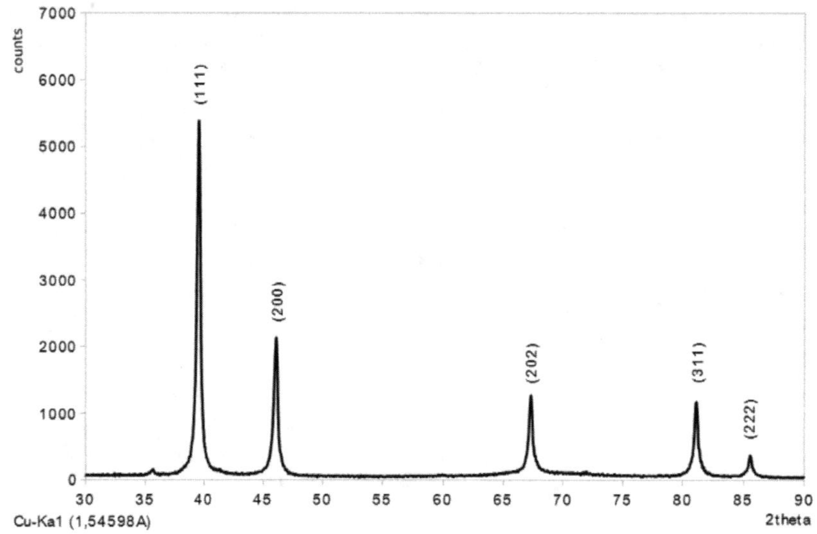

Figure 2. X-ray pattern of platinum nanoparticles.

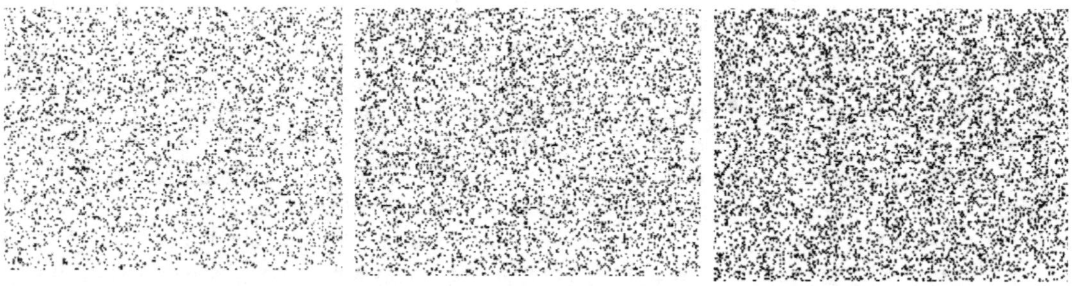

One layer loaded 10,9 %

Two layers loaded 15,0 %

Three layers loaded 21,1 %

Figure 3. Coverage of the surface by nanoparticles of platinum, expressed as a percentage.

IOP Conf. Series: Materials Science and Engineering **706** (2019) 012008 doi:10.1088/1757-899X/706/1/012008

The curves shown in figure 4, 5 and figure 6 showed the charge transfer resistance describes by arcs on graphs and Warburg impedance visible at highest values of ReZ as a straight line at the angle 45° to the axis. It is noticed on graphs that charge transfer resistance value increases over time – the arcs increases.

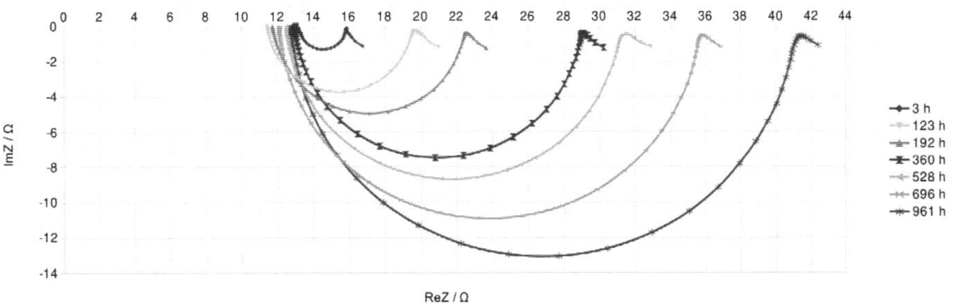

Figure 4. Nyquist plot of cell consist one Pt layer vs. time of electrolyte influence.

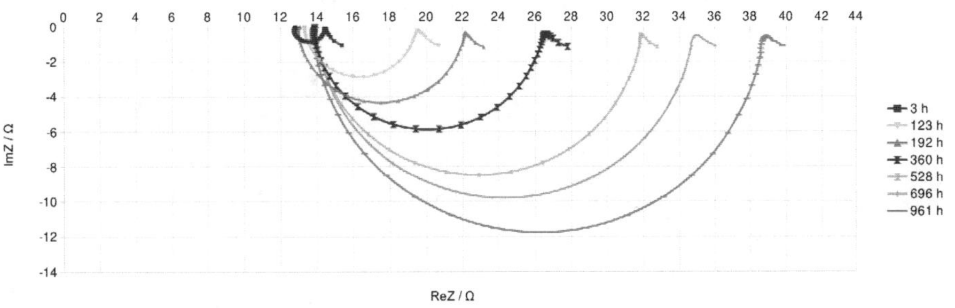

Figure 5. Nyquist plot of cell consist two Pt layer vs. time of electrolyte influence.

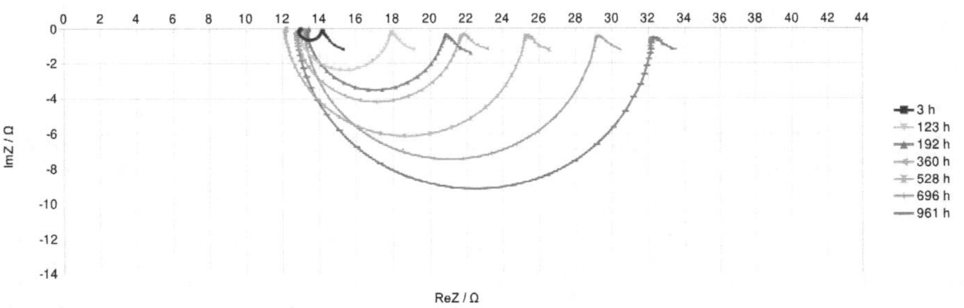

Figure 6. Nyquist plot of cell consist three Pt layer vs. time of electrolyte influence.

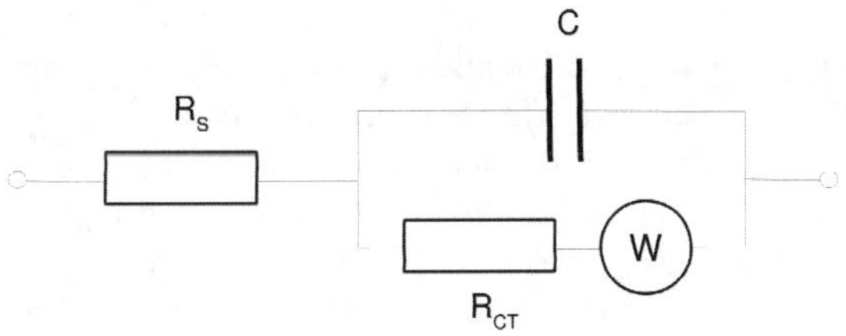

Figure 7. Equivalent circuit for the impedance spectrum in the cell.

A circuit diagram, and its components, is shown in figure 6. The schema consists of a series resistance (Rs) which corresponds, mainly, to the resistance of the electrolyte and the resistance of the connectors. The second element, labelled R_{CT}, is a charge transfer resistance which is convenient for describing the platinum layer properties for DSSC applications. Capacity (C on this schema) characterises the phenomenon that occurs near the electrode surface and the Warburg diffusion element (W) models the diffusion processes that were observed at the lowest frequency.
The values of elements, calculated by dedicated SIM software, versus time are shown in Table 1.

Table 1. Values of schema elements over time calculated for I, II and III layer of platinum.

element	1 – 4 h*			124 h			961 h		
	I	II	III	I	II	III	I	II	III
R_s (Ω)	12.8	12.3	12.1	11.6	13.4	12.9	12.9	13.9	12.9
R_{CT} (Ω)	3.0	1.6	1.2	7.8	6.0	4.9	27.6	24.4	19.1
W (DW)	4.2	3.8	4.3	4.5	4.1	4.3	5.0	4.6	4.3
C (μF)	4.9	4.9	9.0	5.6	7.2	8.3	3.9	4.7	5.9

* mean values calculated for first 4 h

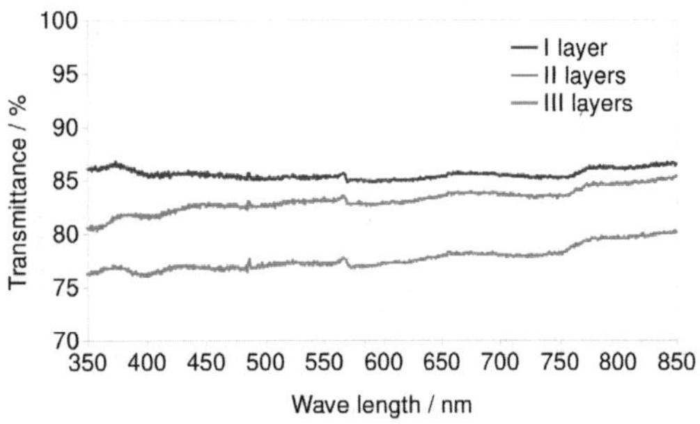

Figure 8. Transmittance of the FTO/Pt counter electrode with different loading of Pt before layers exposure in electrolyte.

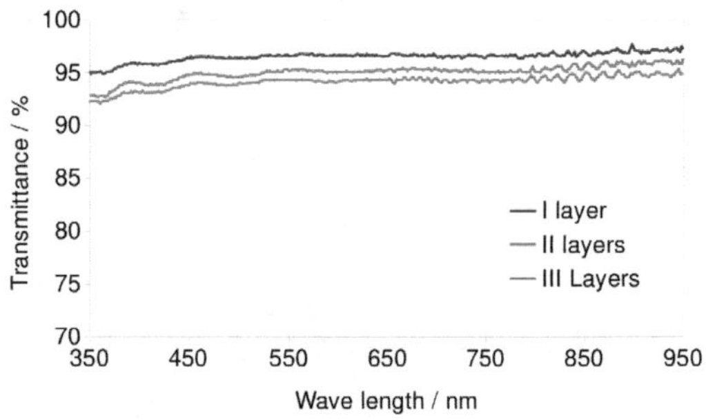

Figure 9. Transmittance of the FTO/Pt counter electrode with different loading of Pt – influence of electrolyte.

It was found that the series resistance, R_s, did not change significantly during the time of the electrolyte contact with the platinum layers in all the tested cells (Table 1). The R_s value oscillated in a range from 12 to 14 Ω, mainly due to the fluctuation of an electrical connections resistance. The mean value of the series resistance, for all the measurements, equal to 12.8 Ω is a result of the conductivity of the applied electrolyte and its ion concentration. Significant changes in R_{CT} values were observed in measurements carried out on each cell. Table 1 shows differences between cells built from I, II or III layers. R_{CT} strongly depends on the loading of platinum on the FTO substrate and the time of electrolyte influence on platinum layer. During the first hours of the iodine/potassium iodide electrolyte influence on the Pt layers the value of charge transfer resistance was stable. Its value was lowest for III layers and highest for I layer (Table 1). The best electrocatalytic properties were obtained for a higher loading of platinum. The mean values of R_{CT}, calculated for the first four hours, were 3.0 Ω, 1.6 Ω and 1.2 Ω, respectively, for I, II and III layers. Further measurements were taken after much longer periods of time to check the stability of the layer properties. An increase of R_{CT} with an increase of time was found. Its value, after 124 h, seems to be too high to apply these layers in solar cells because of possible drop of current result in lower efficiency of solar cell. With an increase in time, the R_{CT} showed much higher values for all tested layers. The cell consisting of III layers was characterised by better properties than the I and II layer cells, but values were not acceptable: I layer – 27.6 Ω, II layers – 24.4 Ω, and III layers – 19.1 Ω, after 961 hours. Compared to the R_{CT} changes, the

IOP Conf. Series: Materials Science and Engineering **706** (2019) 012008 doi:10.1088/1757-899X/706/1/012008

Warburg impedance did not change significantly. This is also shown in figure 4, figure 5 and figure 6. The double layer capacity (C) depends on the platinum loading. The highest capacity was found for a higher amount of Pt with its decreasing over time. The same dependence can be seen for a smaller platinum loading (I layer). The most important changes observed in this study are the changes in the R_{CT} value with increasing time. Using UV/Vis spectroscopy applied for this purpose by other researchers [23] the differences in loading of platinum, before preparing the symmetrical setup, was checked by measuring the transmittance of light through the FTO glass covered by I, II and III layers (figure 8). The highest transmittance (about 85 %) was found for I layer of platinum which is a result of the lowest loading of Pt. The electrocatalytic properties of Pt loaded on an FTO substrate deteriorated with increasing exposure time to iodine/potassium iodide electrolyte. A larger loading of Pt did not resolve the problem. It was found the same phenomenon in all three type of layers (I, II and III) with different amount of Pt. The increases of R_{CT} are a result of the decreasing of nanoparticles number on FTO surface, which in the effect of electrolyte contact influence. That has been confirmed by transmittance measurements. A much higher transmittance of the Pt layers after 961 h exposure in the electrolyte, compared to the same layer before exposure, was found (figure 9). That is a result of decreasing of platinum amount of FTO surface which cause the increases of values of the charge transfer resistance R_{CT}.

4. Conclusions

Platinum counter electrodes, prepared by a spin-coating technique from a low concentrated solution containing nanoparticles of Pt, are characterised by good electrochemical properties, depending on the platinum loading and the time of iodine/potassium iodide electrolyte influence. Reported in this work results showed that described method can be used for preparing counter electrode only for laboratory tests which could be useful for DSSC cells examination. Platinum electrocatalytic layers obtained by this method provide good stability over time during first 4 hours after preparing which is enough for taking measurements. The advantages of this method is possibility of preparing very low concentrated solution of the platinum nanoparticles and use them for layer preparation when it will be needed. The best electrocatalytic activity, for triiodide reduction, was found for a higher platinum loading (III layers) by spin-coating. Charge transfer resistance R_{CT} was satisfactorily low. The influence of the electrolyte on the Pt counter-electrode, results in a significant increase in the R_{CT} value with increasing time. This is a result of the smaller amount of platinum nanoparticles adhering to the FTO, which was illustrated by the transmittance measurements. The same results were found for all the tested loadings. To summarise, counter electrodes can be prepared using the low concentrated Pt precursors in polyol process and the obtained Pt is suitable for testing dye sensitised solar cells because it has good electrocatalytic properties, for a sufficiently long time, to enable exact measurements to be taken.

5. References

[1] Pageorgiou N 2004 *Coord. Chem. Rev.* **248** 1421

[2] Calogero G, Calandra P, Irrera A, Sinopoli A, Citro I and Di Marco G 2011 *Energy Environ. Sci.* **4** 1838

[3] Murakami T N and Grätzel M 2008 *Inorganica Chimica Acta* **361** 3 561

[4] Ehsan M A, Younas M, Rehman A, Altaf M and Khan Y 2019 *Polyhedron* **166** 186

[5] Ye M, Wen X, Wang M, Iocozzia J, Zhang N, Lin C and Lin Z 2015 *Materials Today* **18** 3 155

[6] Feldmann C 2005 *Solid State Sciences* **7** 868

[7] O'Regan B and Grätzel M *Nature* 1991 **353** 6346 737

[8] Kroon J M *et al* 2007 *Prog. Photovolt: Res. Appl.* **15** 1

[9] Hinsch A *et al* 2008 *Prog. Photovolt. Res. Appl.* **16** 6 489

[10] Bonet F, Delmas V, Grugeon S, Urbina R H, Silvert P Y and Tekaiaelhsissen K, 1999 *Nanostructured Materials* **11** 8 1277

[11] Skupien K *et al* 2009 *Prog. Photovolt. Res. Appl.* **17** 1 67

[12] Boukamp B A, Henny N H, Bouwmeester J M and Blank D H 2015 *A Solid State Ionics* **283** 81

[13] Drach Z, Hershkovitz S, Ferrero D, Leone P, Lanzini A and Santarelli M 2016 *Solid State Ionics* **288** 307

[14] Dupuy L *et al* 2009 *Superlattices and Microstructures* **46** 205
[15] Wang Q, Moser J E and Grätzel M 2005 *J Phys Chem B.* **109** (31) 14945
[16] Karim N A, Mehmood U, Zahid H F and Asif T 2019 *Solar Energy* **185** 165
[17] Kim S J, Kwon J, Nam J K, Kim W and Park J H 2017 *Electrochimica Acta* **227** 95
[18] Olsen E, Hagen G and Lindquist S E 2000 *Solar Energy Materials & Solar Cells* **63** 267
[19] Syrrokostas G, Siokou A, Leftheriotis G and Yianoulis P 2012 *Solar Energy Materials & Solar Cells* **103** 119
[20] Kim S S *et al* 2006 *Electrochimica Acta* **51** 3814
[21] Chen L *et al* 2010 *Electrochimica Acta* **55** 3721
[22] Lee K M *et al* 2014 *Electrochimica Acta* **135** 578
[23] Marszalek K, Winkowski P and Jaglarz J 2014 *Material Science Poland* **32** 1 80

IOP Conf. Series: Materials Science and Engineering **706** (2019) 012009 doi:10.1088/1757-899X/706/1/012009

Contribution of polypropylene fibres melting to permeability change in heated concrete – the fibre amount and length effect

I Hager[1], K Mróz[1] and T Tracz[1]

[1] Chair of Building Materials Engineering, Faculty of Civil Engineering, Cracow University of Technology, Cracow, Poland

E-mail: ihager@pk.edu.pl

Abstract. The paper presents the investigations related to concrete permeability change during heating and the contribution of polypropylene fibres melting at the temperature of 163°C to permeability. Permeability is an important parameter governing the occurrence of the explosive behaviour also known as concrete spalling due to fire. To limit the spalling of concrete, the polypropylene fibres are used to act as a fuse and, by melting, reduce gas pressure in concrete pore system. The investigations consisted of testing the permeability change of seven high performance concretes with polypropylene fibres. The permeability was measured after temperature exposure to temperature 140, 160, 180 and 200°C. The results have shown to what extent the fibre length and content affect the permeability when melting. Moreover the results have shown that the addition of polypropylene fibres had the desired impact causing a considerable increase in concrete permeability after fibre melting.

1. Introduction

The concrete infrastructure fires in tunnels (Mont Blanc tunnel, Chunnel tunnel) pointed at the problem of concrete explosive behaviour in fire which is known as fire spalling. Concrete spalling results mostly from thermal gradient within the concrete element, restraint of thermal strains and gas pore pressure increase in the concrete pore system. Low permeability cementitious materials like high-performance concrete (HPC) are characterised by low water-cement ratio and mineral additives incorporation. Those dense materials are particularly prone to spalling. One of the spalling prevention methods consists in the application of fibres that melt in the temperature lower than the temperature of maximum pore pressure occurrence, which is of 180-260°C [1]. Fibres melting temperature depends on the polypropylene (PP) type: for the low melting temperature polypropylene it is of 120°C, and for typical PP fibres is reaching 171°C. The PP fibres melting leads to favourable increase in permeability allowing the water vapour gas release. The study investigated various type of fibres: shorter 9 mm in length and longer 12 and 19 mm in length. In this study, two amounts of fibres were used 0.9 kg/m^3 and 1.8 kg/m^3. This amount corresponds to the amounts that are used commonly [2-4] to prevent spalling occurrence (0.1÷0.2% of concrete volume). Getting close to the melting temperature of fibres, melted polypropylene is absorbed by the cement matrix [1]. That is how a network of open pores, increases permeability, and consequently reduces the internal pressure in the heated concrete. However, the exact mechanism by which PP fibres work is still not fully explained. The increase of micro-cracking in heated concretes with PP fibres was observed. The presence of fibres in the cement matrix may be considered as discontinuity that favours the initiation and development of micro-cracks during heating. On the other hand, fibres are expanding by 10 % while melting what lead to crack development that contributes to an increase in concrete permeability.

2. Materials and methods

Seven high performance concretes were designed with the same constituents apart of the fibre type and its amount. Corse basalt aggregates and Portland cement CEM I 52.5R were used. The concrete water-cement ratio was of 0.3, and it was obtained thanks to the plasticiser and superplasticiser application. The fibres for this research are produced in the fibrillation process consisting of cutting the polypropylene sheets into fibres with lengths of 6, 12 and 19 mm. The PP used in this research has a density of 0.91 g/cm^3. Three dosage levels 0; 0.9 and 1.8 kg/m^3 were used in tested concretes. The fibres melting temperature was of 163°C, and their burn-out temperature 360°C. The compositions of all testes concretes are presented in table 1 while amounts of PP fibres used in research are indicated in table 2 for individual compositions.

Table 1. Mix design of all tested HPCs.

Cement (kg/m^3)	Water (kg/m^3)	Sand 0/2 mm (kg/m^3)	Basalt agg. 2/8 mm (kg/m^3)	Basalt agg. 8/16 mm (kg/m^3)	Plasticizer (%mc)	Super Plasticizer (%mc)	w/c -	Cement paste content (dm^3/m^3)	Mortar content (dm^3/m^3)
490	145	611	712	712	0.90	1.0 ÷ 2.0	0.30	330	514

Table 2. PP fibre dosage for individual tested HPC.

	Length of fibre	HPC	HPC 0.9_6	HPC 1.8_6	HPC 0.9_12	HPC 1.8_12	HPC 0.9_19	HPC 1.8_19
PP fibres (kg/m^3)	6 mm	-	0.9	1.8	-	-	-	-
	12 mm	-	-	-	0.9	1.8	-	-
	19 mm	-	-	-	-	-	0.9	1.8

For permeability tests, cylindrical samples Ø150 mm and 50 mm high were used. The28 days compressive strength was determined using 100 mm cubic samples. All samples were formed and cured in compliance with the PN-EN 12390-2 standard. After the 28 days compressive strength values are presented in the table along with the reference permeability.

The permeability tests were conducted after 90 days of curing and heating in the furnace with a constant rate of heating 1°C/min. Samples were heated from the ambient to the temperatures of 140, 160, 180 and 200°C and maintained at the target temperature for two hours. After cooling down the gas permeability using nitrogen was tested. Permeability was also established for the reference stage on samples dried to constant mass at a temperature of 105°C, according to the guidelines for the method applied described in RILEM Technical Recommendation [5]. Each measurement set comprised of three samples.

3. Results

3.1. Initial values of permeability and compressive strength

The compressive strength f_{c28} and permeability k values are given in table 3. Considering the compressive strength test results, it may be concluded that this property does not vary considerably for tested concretes. Slightly higher initial permeability values for fibres content of 1.8 kg/m^{3}may be explained by the higher porosity of concretes introduced during mixing.

Table 3. Initial values of f_c and permeability of conccrete with and without PP fibres

Label	f_{c28} (MPa)	k (m^2)
HPC	106.5	5.451 x10^{-17}
HPC_0.9_6	110.5	8.097 x10^{-17}
HPC_1.8_6	111.0	7.220 x10^{-17}
HPC_0.9_12	109.8	11.94 x10^{-17}
HPC_1.8_12	108.0	15.00 x10^{-17}
HPC_0.9_19	104.8	10.51 x10^{-17}
HPC_1.8_19	101.0	52.89 x10^{-17}

3.2. Permeability change with temperature – influence of fibres length

It can be observed that an increase in heating temperature results in permeability increase, figure 1. The sharp permeability increase was observed between two measuring points at 140 and 160 °C. A significant increase in residual permeability takes place after heating to 160°C, thus to a temperature of 3 degrees lower than the melting temperature of the fibres used. This observation leads to a conclusion that the fibres are efficient even at temperature lower than melting temperature. This observation is in accordance with the observations of Kalifa [1] who reported the development of micro-cracks around the fibre. Nevertheless, the SEM observations suggest that the increase of permeability at this temperature may be caused by debonding of fibres from the cement paste. In figure 2 the contact zone between the cement paste and the surface of the fibre is presented.

For concrete without fibres (HPC) and concrete with a small addition of short fibres (HPC_0.9_6), this increase is relatively small. However, after heating to 200°C, the permeability increases three times. The addition of 0.9 kg/m^3 of fibres causes a significant increase in permeability for longer 12 and 19 mm fibres. At 160°C, the observed change in permeability reaches 6 and 12 times of the initial value.

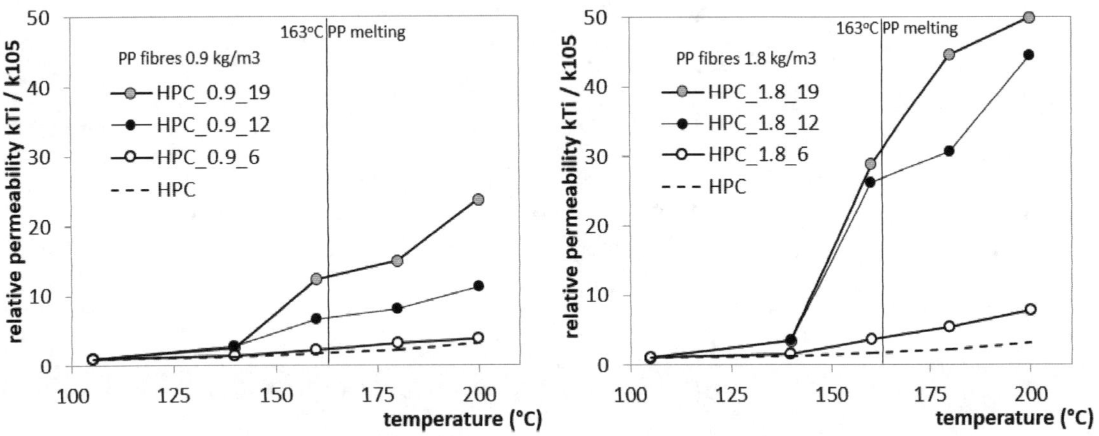

Figure 1. Change of relative permeability with temperature - influence of the length of the fibres.

It can be also concluded that longer fibres after melting are more likely to form a percolated open pore system and contributing to existing porosity and adding to more porous interfacial transition zone (ITZ). The porosity increase of concretes with fibres in an amount of 0.9 kg/m^3 and length of 6; 12 and 19 mm, leads to permeability increase of 3; 11 and 24 times higher than initial. When the amount of fibres was increased to 1.8 kg/m^3, a: 45- and 50-fold increase in permeability was recorded for HPC with 12 and 19 mm long fibres respectively.

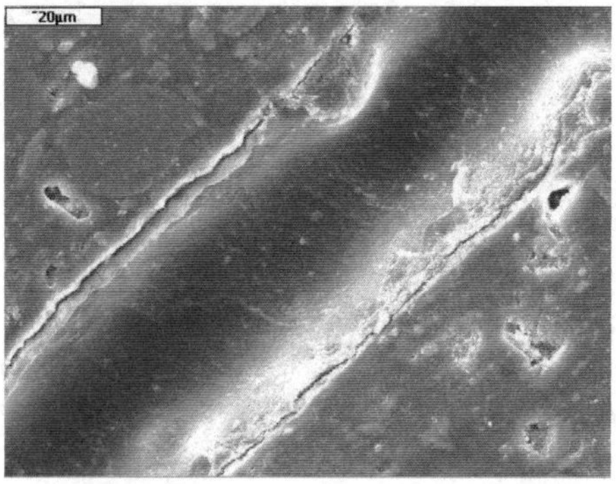

Figure 2.SEM; x1000 PP fibre in concrete heated to temperature bellow melting. The deboning of fibre from the matrix.

Figure 3.SEM; x1000 PP fibre bed after melting and burning out of fibre, heating to 600°C. Visible micro-cracks.

3.3. Permeability change with temperature – influence of fibre content

Based on the results presented in figure 4, it may be concluded that the amount of fibre strongly affects the changes in residual permeability caused by HPC heating. As expected, the highest increase in the residual permeability of heated concretes can be observed in samples in which higher fibre contents 1.8 kg/m^3.

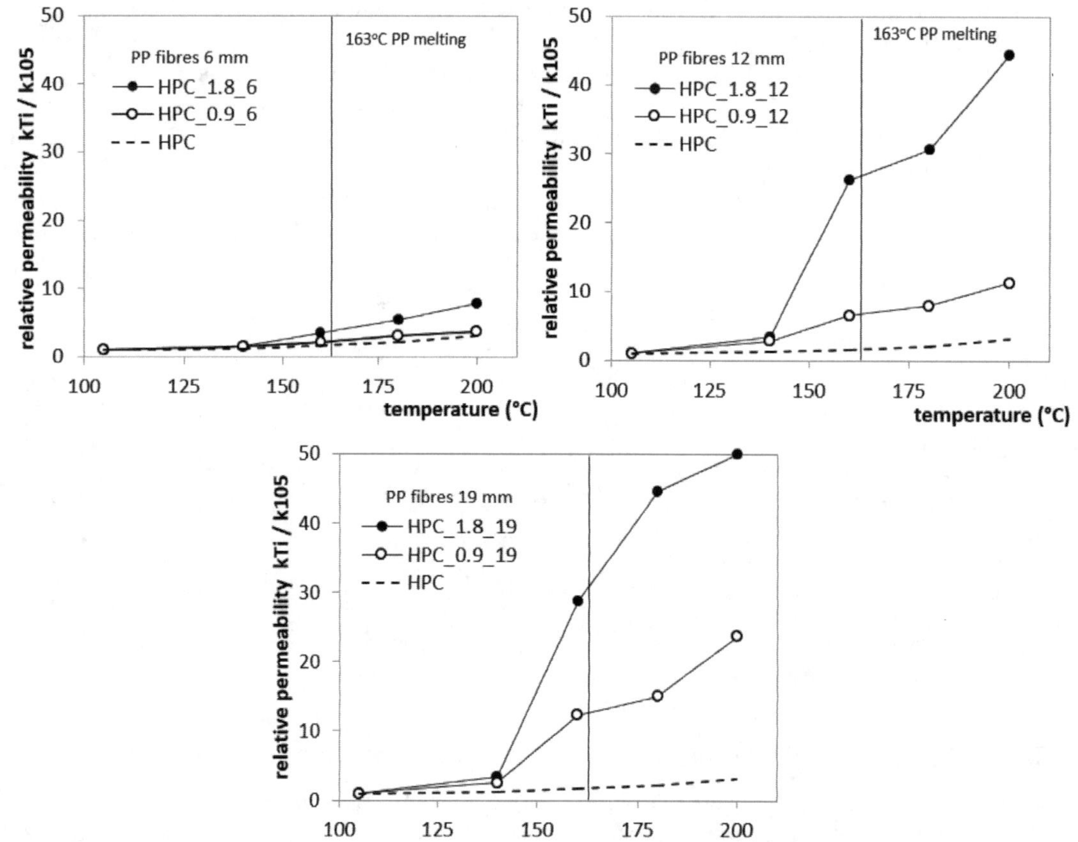

Figure 4. Change of relative permeability with temperature - influence of fibres content.

4. Conclusions
On the basis of the presented results, the following conclusions can be drawn:

- The addition of polypropylene fibres in amounts of 0.9 and 1.8 kg/m^3 does not affect strongly the initial properties of concrete. The compressive strength and permeability values have remained very close for all investigated concretes. Difficulties in homogenisation may have caused slightly higher values of permeability measured for concretes with greater PP fibres dosage.
- Increase in heating temperature results in progressive permeability increase mainly due to polypropylene fibres melting that occurs at 163°C. Nevertheless, the even lower exposure temperature of 160 °C induced substantial permeability increase.
- The sharp permeability increase was observed between two measuring points at 140 and 160 °C. A significant increase in residual permeability takes place after heating to 160°C, thus to a temperature of 3 degrees lower than the melting temperature of the fibres used. The sharp permeability increase after exposure to 160 °C may be caused by partial absorption of fibres by cement matrix and due to debonding of fibres from the cement matrix.
- The length of PP fibres has a significant impact on permeability development. Short fibres seem to be not efficient in creating interconnected and permeable pore network. Although the total dosage of PP fibres was equal, the short fibres of 6 mm provide 4-fold lower permeability than fibres of 12 mm in length and 8-fold lower than 19 mm long fibres. It is concluded that except for increased porosity by melting of added fibres, the fibres melting can also build a connected network of micropores that can facilitate the moisture flow through the material.
- The amount of PP fibre plays a significant role in the development of permeability of concrete subjected to elevated temperature. It was observed that up to 140°C the results for different dosages are quite similar. Over this temperature, the impact of PP fibre dosage starts to differ. As spalling starts most frequently at temperature over 150°C, it is even more important to emphasize the influence of higher amounts of PP fibres on concrete permeability increase.

The increase of concrete permeability in fire situation in case of dense and tight concretes is vital for limiting concrete spalling propensity. The longer polypropylene fibres are proven solution enabling to increase permeability in an effective way. PP fibres provide a supplementary pore network at temperature lower than spalling occurrence, enabling gas pore pressure to be reduced. Therefore, the appropriate concrete mix design is of high importance, especially while designing the high performance concretes concrete structures with the risk of fire occurrence.

Acknowledgements
This research was funded in part from National Science Centre, Poland (NCN) – grant 2016/23/N/ST8/01155 and from EMMAT project E-mobility and sustainable materials and technologies PPI/APM/2018/1/00027 financed by the Polish National Agency for Academic Exchange (NAWA). The author K. Mróz obtained financial resources as part of financing the doctoral scholarship from the National Science Centre, Poland - 2018/28/T/ST8/00352.

5. References
[1] Kalifa P, Chéné G and Gallé C 2001 High-temperature behaviour of HPC with polypropylene fibers: From spalling to microstructure *CemConcr Res* 10 **31** 1487 99
[2] Han Ch-G, Hwang Y-S, Yang S-H and Gowripalan N 2005 Performance of spalling resistance of high performance concrete with polypropylene fiber contents and lateral confinement *Cem Concr Res* **35** 1747–53

[3] Lo Monte F, Felicetti R and Rossino Ch 2015 In-plane loaded concrete slabs subjected to fire: a novel test set-up to investigate spalling *Proc. IFireSS – International Fire Safety Symposium* (Coimbra)

[4] Akasaka H, Ozawa M, Parajuli S S, Sugino Y, Akutsu Y and Murakami M 2018 Preventive Effect on Fire Spalling of High-Strength Concrete With Jute Fibre in Ring-Restraint Specimen *14th International Conference on Concrete Engineering and Technology* (*IOP Conf. Series: Materials Science and Engineering* vol 431*)* 042001

[5] RILEM Technical Recommendation 1999 Tests for gas permeability of concrete, TC 116-PCD: Permeability of concrete as criterion of its durability *Materials and Structures* **32** 174-79

IOP Conf. Series: Materials Science and Engineering **706** (2019) 012010 doi:10.1088/1757-899X/706/1/012010

Strength and leachability of geopolymers with the addition of municipal solid waste ashes

M Łach[1], M Hebdowska-Krupa[1] and N Komar[2]

[1]Institute of Materials Engineering, Faculty of Materials Engineering and Physics, Cracow University of Technology, Jana Pawła II 37, 31-864 Cracow, Poland.
[2] Ekologia Przedsiębiorczość Innowacje Spółka z o.o., Kuhna 17, 42-256 Olsztyn k/Częstochowy, Poland.

E-mail: michal.lach@pk.edu.pl

Abstract. The paper presents the results of the study concerning the possibility of immobilization of secondary waste from municipal waste incineration plants in geopolymers. The study was conducted for 2 different sources of ash from Poland and Lithuania. Three different types of waste were tested. Geopolymers were made with 70% added waste from incineration plants. The samples were made in many variants: geopolymers were made on the basis of fly ash and metakoline and waste was added untreated and after washing with water in a ratio of 1:5 (solid/liquid). Compressive strength and water leaching tests were carried out. As a result of the study, geopolymers as a material for immobilization of secondary waste from waste incineration plants were found to be useful. Water leaching tests proved that all the tested wastes stabilized in geopolymers are characterized by very low leachability values of metals such as: copper, zinc, lead, etc.

1. Introduction

One of the methods of disposal of post-processing waste from incineration plants is geopolymerization, which has been increasingly used in many industrial sectors in recent years [1-4]. Geopolymers are a group of materials defining inorganic aluminosilicate polymers with specific composition and properties. They are hard, mechanically resistant materials which also have extraordinary resistance to chemically aggressive environments. The term "geopolymer" was first used by the French scientist Professor Joseph Davidovits in 1970. These materials were initially developed as a fire-resistant alternative to organic thermosetting polymers. Since then, however, the main applications for geopolymer binders have been in the construction industry [1, 2].

Geopolymeric materials are a family of modern construction materials with a number of properties that enable them to replace popular engineering materials such as traditional concrete or building composites. In the future, they may also become an ideal material for the disposal of hazardous materials, including ashes from waste incineration plants [11-12]. Scientific research shows that in comparison to Portland cement technology, solutions for the containment of hazardous waste in geopolymers and alkaline activated binders offer much greater possibilities for immobilization of heavy metals. Studies on the possibility of metals immobilization in alkali activated binders [5] have shown that alkali activated slags can be used as a means of immobilizing heavy metals. The microstructure of alkaline-activated slags shows a higher number of gel pores compared to Portland cement-based slags and at the same time a much lower number of capillary pores. The use of MSWI bottom ash may also increase the density and homogeneity of geopolymer and cement matrix [6].

The results showed that metal leaching in a geopolymer matrix based on fly ash from incineration plants and volcanic ash was significantly reduced. The leaching levels were much lower than the limit set by the United States Environmental Protection Agency (USEPA) [7].

Although geopolymer technology is more effective for leaching pollution compared to other technologies, the study [8] has shown that the values obtained for e.g. chlorides are still higher than the restrictions imposed by e.g. Italian regulations for the disposal of stabilized waste in landfills for non-hazardous waste. As regards the release of chlorine, the values detected in waste stabilized in geopolymers indicate that pre-treatment of waste is required.

The available scientific literature increasingly describes the need for pre-treatment by waste washing. It is not possible to carry out an effective process of stabilizing fly ash from municipal waste incineration without appropriate pre-treatment [6,9, 10].The pre-washing process is most often used to remove chlorides that have a negative effect on the stabilization of a solidified cement-based matrix [13]. The conducted study [14] allows to state that the preliminary treatment of waste consisting in rinsing with water increases the final strength of geopolymers with waste compared to unwashed ashes. IR spectra also suggest that geopolymers with water-washable ashes can provide a more chemically stable heavy metal capsule.

This article presents the results of the study of compressive strength and water leaching for geopolymers with introduced waste from municipal waste incineration plants. The focus was on the comparison of the leaching values for untreated waste and waste after the rinsing process.

2. Material and methods

2.1. Material

Geopolymers were made in two variants: based on 30% of fly ashes from the Skawina CHP Plant and based on 30% of metakaoline. 70% of processed waste (after washing processes) and unprocessed waste was used as additives/fillers from waste. To produce geopolymers flakes of technical sodium hydroxide were used and an aqueous solution of sodium silicate (R-145) whose molar ratio was 2.5 and density was about 1.45 g/cm^3. The tap water was used instead of the distilled one. The alkaline solution was prepared by means of pouring the aqueous solution of sodium silicate over the solid sodium hydroxide. The solution was thoroughly mixed for 15 minutes in a low-speed mixer and allowed to equilibrate until a constant concentration and temperature. Next, the obtained paste was poured into cubic molds 100x100x100 mm. The solidification was conducted on the vibratory table. Then the molds were heated for 24 hours at 75°, cooled to the ambient temperature, taken out of the molds and stored for 28 days.

Pre-treatment of waste - the washing process - was carried out in the following way - the waste was rinsed with running water in the ratio of 1:5, then the excess water was filtered and dried in a dryer to a constant mass. The rinsing time was 15 minutes. The rinsing was carried out in a slow-running laboratory mixer.

The study was carried out using 3 types of waste from municipal waste incineration plants. Description and marking of waste used are presented in Table 1.

Table 1. Symbols and descriptions of the materials tested.

Symbol	Description of the waste
FY1	The fly ash proceeding from a waste incineration plant in the city of Klaipeda in Lithuania,
FY2	The fly ash from a waste incineration plant in Konin (Poland)19 01 13*
FY3	The fly ash from a waste incineration plant in Konin (Poland)19 01 15*

Table 2 presents variants of prepared samples and their determinations. For example, a sample labeled FFY1W is a fly ash based geopolymer with the addition of FY1 waste from the rinsing process. Geopolymers were prepared in 6 composition variants and for each variant rinsed and untreated wastes were used. In total, 12 different compositions were prepared and subjected to further examination.

Table 2. Labels and descriptions of samples.

Symbol	Geopolymer based on:	Type of waste	Washing
MFY1	Metakaoline	FY1	*NW; **W
MFY2	Metakaoline	FY2	*NW; **W
MFY3	Metakaoline	FY3	*NW; **W
FFY1	Fly ash	FY1	*NW; **W
FFY2	Fly ash	FY2	*NW; **W
FFY3	Fly ash	FY3	*NW; **W

*NW – not washed

**W- washed

2.2. Research methods

Compressive strength tests according to EN 12390-3, using Matest 3000 kN, were conducted on cubic samples 100x100x100 mm conditioned at room temperature for 28 days. Five repetitions were carried out for each type of sample.

Leaching tests were conducted according to the PN-EN 12457-4:2006 standard. The samples were prepared according to the PN-EN ISO 15587-2:2005 standard. Depending on the type of analyzed substance, one of the following methods was used: gravimetry, spectrophotometry, ion chromatography and inductively coupled plasma atomic emission spectroscopy (ICP-OES). The mass of each sample was approximately 2.5 kg.

3. Results and discussion

Table 3 present a summary of compressive strength results for geopolymers with untreated and treated wastes. The highest values of compressive strength are characteristic for geopolymers containing 70% FY1 ash, both based on fly ash (12.3 MPa) and metakaoline (8.9 MPa). It should be noted that high compressive strength values were obtained in this case only for geopolymers containing rinsed waste. In the case where no previous treatment has been carried out, the geopolymers with waste FY1 have a compressive strength below 1.

Geopolymers containing waste from the Konin incineration plant with code 190113* are characterized by compressive strength - 2.9 MPa for waste rinsed with water, based on fly ash, 5.4 MPa for waste rinsed with water, based on metakaoline.

Table 3. Results of compressive strength tests of geopolymers made with the addition of waste from municipal waste incineration plants.

Symbol	Compressive strength	
	Not processed	After washing H_2O
FFY1	<1	12,3
FFY3	4,2	2,1
FFY2	<1	2,9
MFY1	<1	8,9
MFY3	2,4	2,5
MFY2	<1	5,4

Figure 1. Minor efflorescences and discolorations on the surface of a geopolymer with FFY1 and FFY2 waste.

The susceptibility to efflorescence was evaluated visually by observing the surface of samples soaked in water. Sample observations are presented in Fig. 2. Only small efflorescences visible on the surface of geopolymers were observed in the tested samples. It was observed that geopolymers containing waste from washing processes are less prone to efflorescence. It should be emphasized that even if no efflorescence was observed in the conducted study, there is a risk of its occurrence at a later date, e.g. during the use of geopolymer products containing waste.

Table 4 presents the results of water leaching tests of the tested wastes and geopolymers made with the addition of these wastes. For each waste, the leachability of raw waste, the leachability values after the washing process and the leachability values of waste immobilized in geopolymer matrices are presented.

Table 4. Leaching of the examined samples

Analyzed substance	Criteria for admision of hazardous waste	FY1	MFY1W	FY2	FY2W	MFY2W	FFY2W	FFY2	FY3	FY3W	FFY3	FFY3W	MFY3W
		mg/kg	mg/kg	mg/kg	mg/kg	mg/kg	mg/kg	mg/kg	mg/kg	mg/kg	mg/kg	mg/kg	mg/kg
Arsenic (As)	25	<0,5	3,2	<1,0	<0,5	6,1	1,5	1,4	<0,2	<0,1	11,8	17,5	3,8
Barium (Ba)	300	76,5	<0,3	82	9,3	<0,3	0,33	0,58	3	1,2	0,16	0,22	<0,3
Cadmium (Cd)	5	0,25	<0,03	<0,05	<0,02	<0,02	<0,03	<0,03	<0,01	<0,01	<0,05	<0,05	<0,03
Chromium (total amount) (Cr)	70	0,26	3,2	0,5	0,35	2,7	0,85	1,2	2,1	1,8	4,2	2,3	1,3
Copper (Cu)	100	225	<0,1	<0,2	<0,05	0,061	<0,1	<0,1	<0,05	<0,05	<0,10	<0,3	<0,1
Mercury (Hg)	2	<0,01	<0,02	<0,01	<0,01	<0,02	<0,02	<0,01	<0,01	<0,01	<0,01	<0,02	<0,02
Molybdenum (Mo)	30	1,5	2,5	1,6	0,68	1,8	4,9	4,4	1,8	0,89	5,3	4,4	2,1
Nickel (Ni)	40	<0.05	<0,1	<0,05	<0,05	<0,10	<0,05	<0,10	<0,05	<0,05	<0,3	<0,2	<0,1
Lead (Pb)	50	365	18,8	125	4	<0,05	0,19	<0,20	2,1	<0,05	<0,05	<0,1	<0,05
Antymon (Sb)	5	<0,5	0,79	<0,5	<0,3	3,1	1,7	2,6	<0,2	<0,10	3,2	5,6	8
Selen (Se)	7	<0,1	0,79	<0,5	<0,3	1,6	2,7	<1,0	<0,2	<0,10	1,9	2,4	2,1
Cynk (Zn)	200	44,5	<0,5	58	8,8	<0,5	1,8	<0,5	18,5	<0,5	<0,5	7,6	<0,5
Chlorides (Cl^-)	25 000	162 000	18 200	177 000	55 300	1 160	27 000	78700	15 000	1 050	9 930	1 290	25 200
Fluorides (F^-)	500	44	56	51	21,5	57	62	51	34	8,4	100	94	600
Sulfates (SO_4^{2-})	50 000	27 800	68200	8 880	10 200	23 200	50 200	41800	14 200	10 200	31 600	25 700	47 100
Dissolvedorganic carbon (DOC)	1 000	24	400	60	65	400	160	71	26	25	140	350	430
Total dissolved solids (TDS	100 000	360 000	150000	448 000	146 000	101 000	152 000	21900 0	69 900	20 300	84 500	107 000	143000
Chromium (VI)		<1,0	<4,0	<1,0	<1,0	<4,0	<1,0	<4,0	<4,0	<4,0	<5,0	<4,0	<2,0

It was observed that due to immobilization in geopolymers and rinsing processes the amount of leachability of such metals as barium, lead, zinc, chlorides was significantly reduced. It was also noted that in some cases the leaching values of e.g. sulfates or dissolved organic carbon increased. This is related to the process of leaching the components of the geopolymer matrix (metakaoline and fly ash). The conducted study confirmed the validity of the processes of preliminary washing of waste and neutralization of waste in the geopolymer technology.

4. Conclusions

The aim of the study was to determine the basic properties of waste solidified in the geopolymer technology. The most important conclusions, which were formulated on the basis of the conducted study, are described below:

In case of geopolymers containing waste from Klaipeda, only after rinsing treatment it is possible to produce geopolymers of high strength - 12.3 MPa for fly-ash geopolymers and 8.9 MPa for metakaoline-based geopolymers. Geopolymers based on fly-ash containing 70% by weight of waste from the Konin incineration plant (code 190113*) are characterized by compressive strength - 2.9 MPa for waste rinsed with water and 5.4 MPa for waste rinsed with water for geopolymers based on metakaoline. In case of waste from the Konin incineration plant, the water flushing process is necessary to obtain geopolymers with strength above 2 MPa. The stabilization and solidification of waste in geopolymers should be carried out using metakaoline (less frequently fly ash). In order to make it profitable, it is possible to use metakaoline obtained e.g. from calcined coal shales. Water leaching tests proved that all the studied waste stabilized in geopolymers is characterized by very low leaching values of metals such as: copper, zinc, lead, etc.

Acknowledgements
1. This work has been supported by the ERA Net-LAC 2nd Joint Call (http://www.eranet-lac.eu) funded by the National Centre for Research and Development in Poland, within the framework of the grant: *Development of eco-friendly composite materials based on geopolymer matrix and reinforced with waste fibers*.
2. This work has been supported by Intelligent Development Operational Program 2014-2020, 1.1.1: Industrial research and development works carried out by enterprises, funded by the National Centre for Research and Development in Poland, within the framework of the grant: 'Development and implementation of eco-innovative post-process waste stabilization technology' ('Opracowanie i wdrożenie ekoinnowacyjnej technologii stabilizacji odpadów poprocesowych'), grant no. POIR 01.01.01-00-0740 / 16-00.

References

[1] Davidovits J 2015 *Geopolymer Chemistry and Applications* (Saint-Quentin, France: Geopolymer Institute)
[2] Provis L and Jannie S. J. van Deventer 2009 *Geopolymers: structure, processing, properties and industrial applications*, Woodhead Publishing
[3] Łach M, Korniejenko K and Mikuła J 2016 *Procedia Engineer.* **151** 410
[4] Davidovits J *30 Years of Successes and Failures in Geopolymer Applications. Market Trends and Potential Breakthroughs*; Geopolymer 2002 Conference, October 28-29
[5] Deja J 2002 *Cement and Concrete Research* **32** 1971–79
[6] Wongsa A, Boonserm K, Waisurasingha Ch, Sata V and Chindaprasirt P 2017 *Journal of Cleaner Production* **148** 49-59
[7] Tome S, Etoh M A, Etame J and Sanjay K 2018 *Recycling* **3** (4) 50
[8] Ferone C *et al* 2013*Materials* **6** 3420-37
[9] Wang X, Li A and Zhang Z 2016 *Procedia Environmental Sciences* **31** 440-446
[10] Hui-Sheng Shi, Li-Li Kan, 2009 *J. Hazard. Mater.* **164** 750–754
[11] Łach M, Mierzwiński D, Korniejenko K, Mikuła J and Hebda M 2018 *Journal of the Air and Waste Management Association* **68** (11) 1190-97
[12] Mierzwiński D, Łach M, Mikuła J and Goły M 2018 *E3S Web of Conferences* **44** 3
[13] Zheng L, Wang Ch, Wang W, Shi Y and Gao X 2011 *Waste Management* **31** 2 311-317
[14] Yakubua Y *et al* 2018 *International Journal of Scientific & Engineering Research* **9** (6)

IOP Conf. Series: Materials Science and Engineering **706** (2019) 012011 doi:10.1088/1757-899X/706/1/012011

Mechanical properties of geopolymers reinforced with carbon and aramid long fibers

M Łach[1], M Hebdowska-Krupa[1], Dariusz Mierzwiński[1] and Kinga Korniejenko[1]

[1]Institute of Materials Engineering, Faculty of Materials Engineering and Physics, Cracow University of Technology, Jana Pawła II 37, 31-864 Cracow, Poland.

E-mail: michal.lach@pk.edu.pl

Abstract. The paper presents the results of the study of geopolymer composites reinforced with carbon and aramid fibers. Geopolymers were made on the basis of fly ash activated with 10 mole sodium hydroxide solution with an addition of aqueous sodium silicate solution. The fibers were introduced in the form of a roving. Carbon fibers 800 tex and aramid fibers 805 tex were used. Additionally, to compare the influence of the number of fibers on the properties of geopolymer composites for carbon fiber, 1600 tex fiber was also used. The flexural strength of composites made with the fibers was tested. Fiber reinforced geopolymer composites showed a higher value of flexural strength than unreinforced ones. The higher value was noted for carbon fiber reinforced composites (800 tex) compared to aramid fiber roving composites (805 tex). The study also showed that the application of higher fiber mass in the case of carbon fiber roving does not bring the expected increase in mechanical properties of the composites. The use of carbon fiber roving (1600 tex) resulted in a lower flexural strength than in the case of 800 tex roving. Double increase of fiber mass in the composite caused a decrease in strength parameters.

1. Introduction

The subject of geopolymer composites reinforcement has been developing dynamically in recent years, offering interesting research and application perspectives. The addition of fibers is mainly aimed at improving mechanical properties, in particular flexural strength [1, 2], and additionally it reduces the propagation of micro-cracks in the material and increases the resistance to brittle fracture [3]. The fibers can also increase the amount of energy absorbed by the geopolymer before the occurrence of damage [4]. An addition of fibers changes the character of the breakthrough from brittle to more ductile. It also reduces the number of cracks and their dimensions-the width of cracks is limited. General brittle behavior is suppressed in favor of an increased ductility [5]. This makes it possible to maintain material consistency for a longer period of time, which can be crucial, especially in emergency situations, in order to save people inside the structure in the event of a construction disaster. Introducing fibers into the geopolymer matrix creates an extremely interesting material for structural solutions, which stands out from other solutions currently available on the construction market.

Modern reinforcements used in geopolymer composites are usually based on inorganic fibers, such as carbon or glass fibers [6, 7] or synthetic fibers [8]. However, the direction of research is clearly visible, undertaking work on natural fiber reinforcement [9, 10]. This solution is to be an environmentally friendly alternative. It is worth noting, however, that natural fiber reinforcements give lower mechanical properties, which does not always allow for their desired applications [11, 12]. Inorganic fibers usually have higher strength properties and higher repeatability than natural fibers, which allows to obtain a homogeneous material [2, 13].

The fiber content in such composites varies significantly, from only 0.1% by weight for short fibers to 20% for mats or felt [14, 15]. Experiments with a higher number of fibers were also conducted, but they do not improve the strength parameters of the composites [16]. Additionally, increasing fiber content may cause a decrease in mechanical strength [17]. It should also be noted that increasing the number of reinforcement layers is not always associated with the improvement of mechanical properties. Studies of flexural strength of geopolymers reinforced with three layers of textile material showed that the parameters of such a composite did not improve compared to those reinforced with two textile layers [18].

In the presented study, the choice of carbon and aramid fibers was made because of the expected improvement in mechanical properties. It is worth noting that while earlier research works on carbon fibers were conducted in a wide range, there are few publications in the case of aramid fibers as reinforcements of geopolymer composites.

An increasing use of carbon fibers is associated with their numerous advantages: low density, high tensile strength and high Young modulus, high fatigue strength and creep resistance, abrasion resistance, infusibility, high chemical resistance, high dimensional stability, good electrical conductivity, low friction coefficient, vibration damping ability and low absorption of X-rays [19]. The disadvantages of carbon fibers include the tendency to oxidation in the oxygen environment under the influence of high temperatures (in non-oxidative atmospheres, at temperatures up to 2000°C carbon fibers do not lose their properties). It should be noted, however, that carbon fiber oxidation is catalyzed by the alkaline environment [19]. This property makes the geopolymer matrix, having an alkaline reaction, a promising possibility for the production of composites, which can also operate at high temperatures.

Carbon fibers in geopolymer composites were most often used in the form of long fibers [20, 21], fabrics [22], carbon fiber felt [23, 24], as well as the addition of other carbon forms, i.e., graphene [25, 26] graphite [27] or carbon fiber waste used in the production of aviation products [28]. The last three additives were most frequently introduced to geopolymers in the form of particles. The main barrier in the application of these fibers on an industrial scale is their high price. However, it should be noted that in the coming years new carbon fiber production technologies are expected to enter the market and this would be related to the decline in their prices. Studies with the use of short carbon fibers were conducted, inter alia, with the use of a metakaoline matrix with the addition of slag from the Italian company Acciaieria di RubieraSpA, in Casalgrande [29]. The flexural strength of composites with 1.0% weight content of 10 μm diameter and 7 mm length fibers was studied. The work carried out showed the reinforcement of the material. For a pure matrix, the flexural strength was 6.9 MPa and for composites with reinforcement it was 11.7 MPa [29].

Research works on carbon fiber reinforced geopolymer composites were also conducted on a geopolymer matrix based on fly ash from the Mae Moh power plant in Lampang, Thailand [30]. Cut carbon fibers in the following weight proportions were added to the composites: 0.0, 0.1, 0.2, 0.3, 0.4 and 0.5%. Two types of specimens were examined-seasoned at 25°C and 60°C. The study was carried out after 7, 14 and 28 days, gradually observing an increase in compression strength of the composites [30]. After 28 days, the best results were obtained for 0.5% of carbon fiber content. For the samples matured at 25°C it was about 43 MPa for the composite with 0.5% fiber added, compared to matrix material that obtained 30 MPa. For the samples matured at 60°C it was about 58 MPa for the composite with 0.5% fiber added. The matrix material produced under the same conditions obtained 45 MPa [30]. Apart from the studies of compressive strength, the research related to electrical conductivity of the composites were also carried out. An increase in conductivity along with an increase in the number of carbon fibers was noted [30].

In the range of ambient temperatures, the main purpose of carbon fibers addition was to increase mechanical properties and change the behavior of composites in the range of fracture characteristics-to avoid brittle breakthrough [6]. However, it is worth noting that these are not the only advantages associated with the use of this fiber-a large part of the research was devoted to the possibility of application of geopolymer composites reinforced with carbon fibers as a material for applications at elevated temperatures. The studies of behavior of geopolymer composites with addition of short carbon fibers at high temperatures were conducted on a geopolymer matrix based on fly ash from the Gladstone power plant in Queensland, Australia [31].The fibers were added in amounts of 0, 0.5, 1 and 1.5% by weight. The study was carried out at temperatures: 28°C, 200°C, 400°C, 600°C and 800°C. The fibers had the following dimensions: diameter about 11 μm and length 6 mm [31]. The results of examination of compression strength at ambient temperature show an increase in its value for composites containing 1 and 1.5% of carbon fiber (about 31 and 32 MPa, respectively) and a decrease in its value for material containing 0.5% of fiber - about 27 MPa, in relation to material without additives-about 29 MPa [31]. At temperatures of 200°C, 400°C and 600°C, the properties of composites increase compared to the corresponding materials tested at ambient temperature-28°C. The properties of some composites decrease only at the temperature of 800°C, while for the specimens with 0 and 1.5% contents the compression strength increases slightly, while for the composites with 0.5 and 1.5% contents it decreases. The highest value in the study was achieved by fiber composites at 200°C, i.e., 0.5%-about 36 MPa, 1%-about 40 MPa and 1.5%-about 36 MPa, respectively. The specimen without the addition of fibers obtained the highest value of compression strength at the temperature of about 37 MPa [31]. The study confirms the possibility of using the composites with the addition of carbon fiber in applications for high temperatures.

Studies on carbon fiber addition, including the influence of temperature on the mechanical properties of composites, were also conducted on a matrix based on a mixture of metakaoline with fly ash. Fly ash came from the Guangzhou Huangpu power plant in China [32]. Carbon fibers of the following dimensions were added to the composites: length: 6 mm, diameter: 7 μm. The following addition of fibers by weight was applied: 0, 0.5, 1 and 2%. The specimens were examined after 7 days at ambient temperature and 500°C [193]. The results of compressive strength at ambient temperature showed a decrease in its value from about 50 MPa for a material without fiber addition to about 45 MPa for a material with 2% addition of carbon fibers. However, at 500°C the material with fibers addition was more resistant and reached about 5 MPa for fiber composites, compared to about 2 MPa for material without fibers [32]. The flexural strength increased with the addition of fibers both at ambient temperature and at 500°C. For the material without fiber addition it was about 5.5 MPa and below 0.1 MPa, while for the composites with 2% addition of carbon fibers it was 15 MPa and 1 MPa [32]. The study showed that the mechanical properties of fiber composites significantly decrease above 200°C [32].

The studies were also conducted on the addition of micro and nano carbon fibers (in the form of nanotubes) to geopolymer composites. The study on the addition of microfibers (fibers of the length of about 100 μm) were carried out in the ratio of 0, 5, 10 and 15% by weight to a geopolymer matrix based on metakaoline [33]. The examinations were carried out after 28 days at temperatures of 30°C, 200°C, 400°C and 800°C. The highest values for temperatures 30°C and 200°C were obtained for 10% microfiber addition, while the values for 200°C were higher than for temperature of 30°C. The values for 30°C were 44.2 MPa for the material with 10% microfibers and 28.4 MPa for the matrix material, while for the temperature 200°C it was 48.8 MPa for the material with microfibers and 36.6 MPa for the matrix material. The highest values for temperatures of 400°C and 800°C were obtained for 15% fiber addition: 33.5 MPa and 24 MPa, respectively, for the same temperatures the compressive strength of the material without fiber addition was: 14.8 MPa and 11.2 MPa [33]. The studies were also carried out on the addition of carbon nanotubes to geopolymers both as an additive to matrices based on metakaoline [34, 35] and fly ash [36]. The addition of nanotubes increased the mechanical properties of composites.

Few studies were carried out with the use of aramid fiber as a reinforcement of geopolymer composites, however, they show a positive influence on the mechanical properties of the composites. The study was carried out with the use of a geopolymer matrix based on fly ash. After the specimens were prepared, they were subjected to high temperature, 85°C for 10 hours [37]. Aramid fibers of 30 mm length and 0.5 mm diameter in 1.0% volumetric ratio were used as an addition [37]. In this case, the matrix material reached the value of compression strength of 70 MPa, while the composites with added fibers-88.0 MPa [37]. Significant increase of the value was observed for flexural strength and it was 10.4 MPa for composite with 1.0% fiber content, compared to 7.1 MPa obtained for pure matrix material and tensile strength increase from 3.1 MPa for matrix material to 7.7 MPa for composite material [37].

The paper presents the results of the study on geopolymeric composites reinforced with carbon and aramid fibers. The fibers were introduced in the form of roving. Carbon fibers 800 tex and aramid fibers 805 tex were used. Additionally, in order to compare the influence of the number of fibers on the properties of geopolymer composites for carbon fiber, the 1600 tex fiber was also used.

2. Material and methods

2.1. Material

Geopolymers were made from fly ash from the CHP plant in Skawina (Poland) and sand in ratio 1:1. The fly ash came from the coal power plant 'Skawina' (located in: Skawina, Lesser Poland, Poland) was used. This fly ash is obtained from a bituminous coal by electrostatic precipitation of fine particles from the exhaust gases from coal-fired furnaces. The chemical composition of the fly ash is typical for class Faccording to ASTM678 standards (ASTM-C618-2). It is rich in oxides such as SiO_2 (55.89%) and Al_2O_3 (23.49%) and contains low amount of CaO (2.72%). It has also good physical parameters - the content of particle under the size 45 μm is about 88%. The morphology of the particles of this fly ash is typical of such by-products of coal combustion and suitable for the process of alkali-activation [15].The silica sand has been use as an aggregate.

Two types of fibers in the form of roving were used: 800 tex carbon and 1600 tex and 805 tex aramid fibers. Selected parameters of the fibers used are given in Table 1.

Table 1. Selected parameters of roving used in the study.

Type of roving	Nominal linear mass of roving, tex	Width	Producer
Carbon	800	5 mm	Havel Composites
Carbon	1600	10 mm	Havel Composites
Aramid	805	8 mm	Havel Composites

The fibers are used in the condition as received from the manufacturer. They were not subject to prior processing.

2.2. Samples preparation

Four series of composites were prepared: with carbon fibres (tex 800) and (tex 1600), aramid fibres (tex 805) and one control series without any fibres. The matrix was prepared using sodium promoter, fly ash, sand. The process of activation has been made by 10M sodium hydroxide solution combined with the sodium silicate solution (the ratio of liquid glass - 1:2) and the tap water. The solution was mixed and left until its temperature became stable and the concentrations equalized, which took, about 2 hours. The fly ash, sand and alkaline solution were mixed about 10 minutes by using low speed mixing machine (to receive the homogeneous paste).

Geopolymer composites were made in molds with dimensions of 20×50×200 mm. Three types of compositions were made in which 1 linear meter of roving, evenly distributed in one layer, and control specimens without reinforcement were used. Figure 1 shows the distribution of fibers in the specimens.

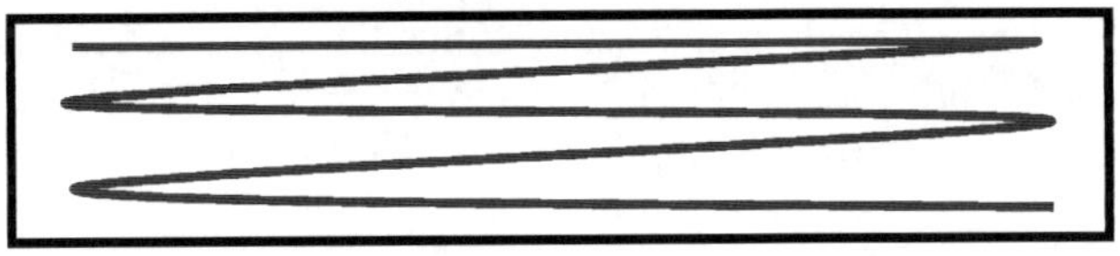

Figure 1. The method of roving arrangement in the specimens prepared for the study-top-down projection.

Next, tightly closed molds were heated in the laboratory dryer for 24h at 75 °C. Then, the samples were unmolded. Each series of samples were tested on flexural strength at ambient temperature after 28days.

2.3. Research methods

Flexural strength tests were carried out according to the methodology described in the standard EN 12390-5 ('Testing hardened concrete. Flexural strength of test specimens'), because of the lack of separate standards for geopolymer materials. The tests for each series involved 5prismatic samples: 50 x 50 x 200 mm (space between supporting points 150mm). Tests were performed on an universal testing machine - MATEST 3000kN with speed 0,05 MPa/s.The calculations werebased on following equation:

$$f_{cf} = \frac{3 \cdot F \cdot l}{2 \cdot d_1 \cdot d_2^2} \qquad (1)$$

where:
f_{cf} – compressive strength, MPa
F – maximal load, N
l – space between supporting points, mm (for conducted tests:140 mm)
d_1, d_2 – sample dimensions, mm

3. Results and discussion

The results from the flexural strength tests are shown in table 2. The results of both methods and samples with fiber reinforced and control specimens (without fibers) are presented.

Table 2. Results of flexural strength testing of geopolymer composites.

Series	Type of fiber	Flexural strenght [MPa]	Average value[MPa]	Standard deviation
1	no reinforcement	7,141	7,2	0,21
		6,920		
		7,218		
		7,440		
2	Carbon fiber	7,930	6,6	0,81
	1mfiber 1600tex	7,021		
		6,014		
		6,588		
3	Carbon fiber	7,861	8,3	0,37
	1mfiber	8,435		
	800tex	8,672		
		8,043		
4	Aramid fiber	8,441	7,7	0,67
	1mfiber	6,930		
	805tex	8,094		
		7,433		

IOP Conf. Series: Materials Science and Engineering **706** (2019) 012011 doi:10.1088/1757-899X/706/1/012011

In the case of geopolymer composites reinforced with fibers of about 800 tex, higher value of flexural strength compared to unreinforced geopolymer was noted. For carbon fiber (800 tex), it was an increase of about 15%, whereas for aramid fiber roving composites (805 tex) it was about 6% compared to the control specimens. The study also showed that the use of higher fiber mass in the case of carbon fiber roving does not bring the expected increase in flexural strength. The use of carbon fiber roving (1600 tex) resulted in a lower flexural strength than in the case of 800 tex roving and a lower flexural strength by about 8% compared to unreinforced material. This may indicate a lack of consistency between the thicker fiber and the matrix. However, this subject requires additional research.

Figure 2. Specimens containing reinforcement in the form of carbon roving 1600 tex.

In relation to the previously analyzed literature studies, it is worth noting that the results obtained with the use of the proposed reinforcement were lower than expected. The previous analysis showed that as a result of carbon fiber reinforcement, the flexural strength improved by approx. 25–55% [29, 33], while in the case of aramid fiber application by 46% [37]. This indicates the need for further work on the reinforcement, or possibly the need to re-engineer its system and/or change the form of the reinforcement, e.g. to short or mixed fibers. It is also worth noting that a significant limitation in this respect is the planned use of composites for the production of incremental technology, which makes it impossible to use reinforcement in the form of mat or fabric.

It should also be noted that neither carbon nor aramid fibers have yet been investigated in additive manufacturing applications, i.e., 3D printing of geopolymers. So far, the only research activities in this respect have not been carried out in the area of steel and glass fibers. These fibers were examined both as continuous reinforcement-steel fibers [38] and dispersed reinforcement-glass fibers [39]. The dispersed reinforcement was added to geopolymer filaments based on fly ash, slag, silica and sand as 3 mm, 6 mm and 8 mm fibers in amounts ranging from 0.25% to 1%. The 1% addition of reinforcement significantly improved the properties of printed materials, especially flexural strength, which facilitated the entire 3D printing process [39].

4. Conclusions

Carbon and aramid fibers in the form of roving were used in the study. The matrix of the composite was a geopolymer made on the basis of fly ash and building sand, activated with 10 M NaOH with the addition of an aqueous solution of sodium silicate in a ratio of 1:2 wt. %. Fiber reinforced geopolymer composites showed a higher value of flexural strength compared to unreinforced geopolymer. The higher value was observed for carbon fiber reinforced composites (800 tex) compared to aramid fiber roving composites (805 tex). The study also showed that the use of higher fiber mass in carbon fiber roving does not bring the expected increase in flexural strength. The use of carbon fiber roving (1600 tex) resulted in a lower flexural strength than 800 tex roving. Double increase in fiber mass in the composite caused a decrease in strength parameters.

The conducted study is an initial research aimed at selecting the most beneficial form of geopolymer composites reinforcement, which are planned to be carried out with the use of additive technology. It is planned to continue research in this area, in particular to determine the influence of fibers on the rheological properties of the composite and to test the influence of fiber addition in terms of providing the composite with other properties, i.e., resistance to temperature or corrosive environments.

Acknowledgements

1. This research was funded in part from National Science Centre, Poland (NCN) – grant 2016/23/N/ST8/01155 and from EMMAT project E-mobility and sustainable materials and technologies PPI/APM/2018/1/00027 financed by the Polish National Agency for Academic Exchange (NAWA).
2. This work has been supported by Smart Growth Operational Programme 2014-2020, IV Increasing the research potential, 4.1.4: Application projects, funded by the National Centre for Research and Development in Poland, within the framework of the grant: *Development of 3D printing technology for construction and facade prefabricated elements made of concrete composites and geopolymers*, grant no. POIR.04.01.04-00-0096/18.
3. This work has been supported by the ERANet-LAC 2nd Joint Call (http://www.eranet-lac.eu) funded by the National Centre for Research and Development in Poland, within the framework of the grant: *Development of eco-friendly composite materials based on geopolymer matrix and reinforced with waste fibers*.

References

[1] Silva FJ and Thaumaturgo C 2003*Fatigue & Fracture of Eng. Mat. & Structures* **26** (2) 167
[2] Mikuła J and Korniejenko K 2015 Innovative, Cost-Effective and Eco-friendly Fibre-based Materials for the Construction Industry (Kraków: Wydawnictwo Politechniki Krakowskiej)
[3] Rickard WDA, VickersL and van Riessen A 2013 *Appl. Clay Sci.* **73** 71
[4] Chen R, Ahmari S and Zhang L 2014*J. Mat. Sci.* **49** 2548
[5] Sakulich AR 2011 *Sustain. Cities Soc.* **1** 195
[6] Lin T, Jia C, He P and Wang M, 2010 *Mat. Sci, Eng. A-Struct.* **527** 2404
[7] Lin T, Jia D, Wang M, He P and Liang D 2009 *B. Mater. Sci.* **32** (1) 77
[8] Yan L, Kasal B and Huang L 2016 *Compos. Part B-Eng* **92** 94
[9] Rashad A M 2018 *Natural Resources Conservation & Research*" **1** 9
[10] Sa Ribeiro R A, Sa Ribeiro M G and Kriven W M 2017 *J. Ceramics Sci. & Technolog y* **8** (3) 307
[11] Korniejenko K, Frączek E, Pytlak E and Adamski 2016 *Procedia Engineer.* **151** 388
[12] Szechyńska-Hebda M, Marczyk J, Ziejewska C, Hordyńska N, Mikuła J and Hebda M 2019 *Materials* **12** (18) 2999
[13] Trindade A, Borges P and Silva F 2019*Advances in Civil Eng. Materials* 8 (article in press)
[14] Davidovits J 2016 *Reinforced Geopolymer Composites: A critical review*, published on line: https://www.materialstoday.com/polymers-soft-materials/features/reinforced-geopolymer-composites-a-critical-review/ (access: 23-09-2019)
[15] Korniejenko K 2019 *Wpływ dodatku włókien krótkich na właściwości mechaniczne kompozytów na osnowie geopolimerowej* (Kraków: Biblioteka Cyfrowa Politechniki Krakowskiej)
[16] Furtos G, Silaghi-Dumitrescu L, Pascuta P, Sarosi C and Korniejenko K 2019 *J. Nat. Fibers*(article in press)
[17] Payakaniti P, Pinitsoonthorn S, Thongbai P, Amornkitbamrung V and Chindaprasirt P 2018 *Materials Today: Proceedings* **5** (6) 14017
[18] Chi H L, Louda P, Periyasamy A P, Bakalova T and Kovacic V 2018 *Fibers* **6** 87
[19] Mayer P and Kaczmar J W 2008 *Tworzywa Sztuczne i Chemia* **6/2008** 52
[20] Pernica D, Reis P N B, Ferreira J A M and Louda P 2010 *J. Mater. Sci* .**45** 744

[21] Tran D H, Kroisová D, Louda P, Bortnovsky O and Bezucha P 2009 *J. Achievments in Materials & Manufacturing Eng.* **37** (2) 492
[22] Zhang H, Hao X and Fan W 2016 *Procedia Engineer.* **135** 47
[23] Yan S, He P, Jia D, Yang Z, Duan X, Wang S and Zhou Y 2016*Ceram. Int.* **42** 7837
[24] He P, Jia L, Ma G, Wang R, Yuan J, Duan X, Yang Z and Jia D 2018*Ceram. Int.* **44** 10726
[25] Yan S, He P, Jia D, Yang Z, Duan X, Wang S and Zhou Y 2015 *Ceram. Int.* **41** (9) 11242
[26] Ranjbar N, Mehrali M, Mehrali M, Alengarama U J and Jumaat M Z 2015 *Cement Concrete Res.* **76** 222
[27] Zhang Y, He P, Yuan J, Yang C, Jia D and Zhou Y 2017 *Ceram. Int.* **43** 2325
[28] Luna-Galiano Y, Leiva C, Villegas R, Arroyo F, Vilches L and Fernández-Pereira C 2018 *Mater. Lett.* **233** 1
[29] Natali A, Manzi S, Bignozzi MC 2011 *Procedia Engineer.* **21** 1124
[30] Payakaniti P, Pinitsoontorn S, Thongbai P, Amornkitbamrung V and Chindaprasirt P 2017 *Constr. Build. Mater.* **135** 164
[31] Shaikh F and Haque S 2018 *Int. J. Concr. Struct. M.* **12** (35) 12
[32] Zhang H, Kodur V, Cao L and Qi S 2014 *Procedia Engineer.* **71** 153
[33] Behera P, Baheti V, Militky J and Naeem S 2018 *Constr. Build. Mater.* **160** 733
[34] Abbasi S M, Ahmadi H, Khalaj G and Ghasemi B 2016*Ceram. Int.* **42** 15171
[35] Yuan J, He P, Jia D, Fu S, Zhang Y, Liu X, Cai D, Yang Z, Duan X, Wang S and Zhou Y 2017 *J. Eur. Ceram. Soc.* **37** 2219
[36] Saafi M, Andrew K, Tang P L, McGhon D, Taylor S, Rahman M, Yang S and Zhou X 2013 *Constr. Build. Mater.* **49** 46
[37] Zhu J, Zheng W Z, Qin C Z, Xu Z Z and Wu Y Q 2018 *IOP Conf. Series: Materials Sci. & Eng.* **292** 012060
[38] Lim JH, Panda B and Pham Q-C 2018 *Constr. Build. Mater.* **178** 32
[39] Panda B, Paul S C and Tan M J 2017 *Mater. Lett.* **209** 146

IOP Conf. Series: Materials Science and Engineering **706** (2019) 012012 doi:10.1088/1757-899X/706/1/012012

Autoclaving of alkali-activated materials

D Mierzwiński[1] and J Walter [1]

[1] Institute of Materials Engineering, Faculty of Materials Engineering and Physics, Cracow University of Technology, al. Jana Pawła II 37, 31-864 Cracow, Poland

E-mail: dariusz.mierzwinski@pk.edu.pl

Abstract. Alkali-activated materials are amorphous materials that look like natural stone or a concrete structure. As in the case of concrete, on the surface of alkali-activated materials, a white coating appears in the form of efflorescence. For conventional efflorescent concrete, the reaction between $Ca(OH)_2$ with water and CO_2 is generally harmless, except for the discolouration. In the case of alkali-activated materials, efflorescence can be a big problem, especially if they are exposed to moisture or in contact with water. The white efflorescence on the surface of the materials is sodium carbonate heptahydrate $Na_2CO_3{\cdot}7H_2O$ [1]. Various chemicals to prevent primary and secondary efflorescence can be used to reduce the appearance of the efflorescence. In the case of alkali-activated materials, this is a more complex process. The following paper proposes the use of hydrothermal processing during alkali-activated materials production. Hydrothermal treatment in an autoclave significantly reduced the appearance of efflorescence on the alkali-activated materials surfaces.

1. Introduction

The use of alkali activated fly ash materials is still low compared to portland cement concrete. One of the reasons is the formation of efflorescence on the surface of the materials. For conventional concrete, efflorescence involves the reaction of $Ca(OH)_2$ with water and CO_2 and is generally harmless. The only harmless phenomenon of such a reaction is discoloration. The intensity of carbonisation of concrete freely exposed to the external environment is greater, the greater the porosity of the concrete. the greater its susceptibility to scratches and cracks (the lower the density of the concrete mix, the greater the w/c ratio in the mix, the worse the concrete care) and the CO_2 concentration in the ambient air is higher [2, 3, 4, 5].

Polish Standards for concrete products (eg PN-EN 1338: 2005 - "Concrete paving stones. requirements and test methods") and the required Technical Approvals of the Road and Bridge Research Institute allow for the occurrence of limestone efflorescence as a natural process of concrete maturation.

Commercial products are used in the form of liquid additions to a concrete mix to ensure watertight concrete surfaces. They prevent solids from absorbing capillary water. They do not affect the hydration of the cement and the hardening process of concrete, because they only become active after the setting process is completed. Since the distance between the cement grains coated with blocking preparations is very small, the surface tension of the water particles does not allow it to penetrate into the intermediate spaces. Surface preparations are also used to help to reduce efflorescence. Preparations based on polymer and silicon-lithium compounds with the highest silicate modules, thanks to penetration and hardening of concrete in the molecular structure, enables the production of a high-strength cement matrix with extreme physical and chemical properties. Additional modification

IOP Conf. Series: Materials Science and Engineering **706** (2019) 012012 doi:10.1088/1757-899X/706/1/012012

with a selective polymer binder ensures maximum sealing of the concrete by binding the concrete matrix components such that they do not contain free calcium compounds. Preparations based on lithium silicate deeply penetrate the concrete layer, reacting with free calcium hydroxide to create a hard, durable, tight and compact structure.

In the case of relatively modified materials, efflorescence can be a big problem, especially when they are exposed to moisture or contact with water. Treatments aimed at limiting efflorescence on the surface of geopolymers have been developed. These include, for example, the addition of alumina or a hydrothermal cure. The addition of alumina significantly reduces the mobility of alkali. However, reduction of efflorescence by means of hydrothermal curing can be achieved by using a curing temperature above 65°C and increasing the humidity during the process [1, 6].

2. Material and research methods

To reduce the amount of efflorescence on the surface of the elements, samples in the form of cubes were prepared in accordance with the EN 12390-1 standard for concrete. The shape and size of the samples were also selected such that tests on the characteristic compressive strength could be carried out after the hydrothermal treatment of the alkali modified materials.

2.1. The material used to make the samples

Fly ash from the Polish power plant in Skawina was used to prepare the samples. The oxide composition of the fly ash is shown in Table 1 and its morphology in Figure 1.

The mass of the samples was prepared from a mixture of dry ingredients, which included, in addition to volatile dust, sand in a 1: 1 ratio and a liquid component, which was a twelve molar (12M) alkaline solution based on sodium hydroxide and water glass. The mixture consistency was such that the ratio of fly ash to alkali was 0.4.

Table 1. Oxide composition of fly ash.

	SiO_2	Al_2O_3	Na_2O	CaO	Fe_2O_3	MgO	K_2O	SO_3	TiO_2	P_2O_5	BaO
[% mas]	55.89	23.49	0.59	2.72	5.92	2.61	3.55	0.16	1.09	0.82	0.20

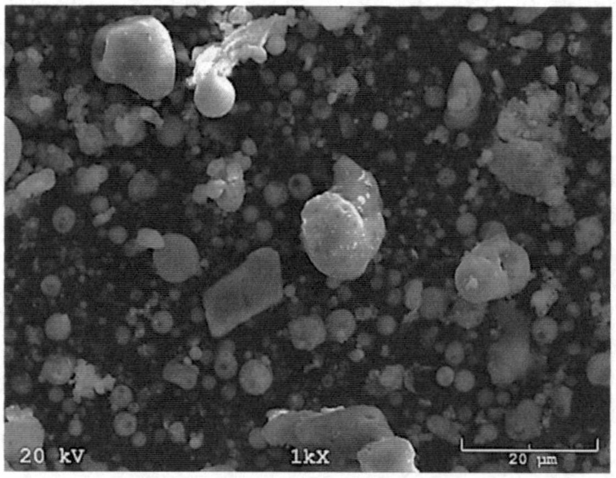

Figure 1. Fly ash morphology.

2.2. Sample preparation

The prepared mass was mixed in a mechanical mixer and then placed in moulds that were subjected to a vibrating process on a vibrating table to remove air bubbles and uniformly fill the moulds. The mixing process lasted about 15 minutes, which had a significant impact on the dissolution of the fly dust particles.

IOP Conf. Series: Materials Science and Engineering **706** (2019) 012012 doi:10.1088/1757-899X/706/1/012012

Using the above method, three sets of samples were prepared, three samples for each hydrothermal treatment method. The samples were marked as follows: SA, SB, SC.

2.3. Autoclaving
The hydrothermal treatment was carried out in a Parr® pressure chemical reactor (autoclave).

A hydrothermal treatment was carried out in an autoclave for two sets of samples.

The SA sample was subjected to an annealing process in a laboratory dryer at a temperature of about 75°C for 24 hours. The purpose of this treatment was to determine the value limits associated with efflorescence and other parameters of the tested material.

The specimen with the SB determination was, after annealing according to the scheme for SA sample, subjected to hydrothermal treatment in an autoclave. The hydrothermal treatment in the autoclave lasted 6 hours at a temperature of 160°C with a superheated steam pressure of 9 bar.

The SC sample was subjected to an annealing process at a temperature of about 75°C for 3 hours in a laboratory drier and then was subjected to a hydrothermal treatment process in an autoclave for 6 hours at 160°C with a superheated steam pressure of 9 bar.

The hydrothermal treatment process took place at a humidity resulting from the temperature and pressure of the superheated steam. The temperature and time of the process were continuously monitored through software dedicated to the Parr® autoclave.

2.4. Compressive strength test
The compressive strength test was performed on cubic samples of 100 x 100 x 100 mm.

The results of compressive strength were interpreted in accordance with PN-EN 13791 [7]. The number of samples in each series did not exceed 4. The value of the characteristic compressive strength is the smaller of the two values, according to formula (1).

$$f_{ck,is} = min\begin{Bmatrix} f_{cm(n),is} - k \\ f_{is,lowest} + 4 \end{Bmatrix} \quad (1)$$

2.5. pH test
The leachate pH test was carried out 14 days after the samples were demoulded. For the first 7 days the samples were seasoned, and for the next 7 days they were immersed in distilled water. The pH of the effluent was measured using a CP-105 type pH meter.

2.6. SEM EDS study
Morphology analysis, after geopolymerisation, was performed on a scanning electron microscope of the JEOL JSM-820 company using the EDS microanalyser. Samples for SEM tests were sputtered with a thin layer of gold on a JEOL JEE-4X vacuum sputtering machine.

3. Results

3.1. Change of sample mass
In order to determine the effect of the hydrothermal treatment after the process of alkaline activation of the materials, a measurement of the change in sample mass was carried out, the results of which are presented in Table 2. Based on the results obtained, a slight increase in the mass of samples after hydrothermal treatment can be observed. The increased mass was also maintained after 28 days of ripening at 21°C and humidity of about 50%, which may indicate a reduced porosity of the sample surface and, consequently, a reduced evaporation surface.

IOP Conf. Series: Materials Science and Engineering **706** (2019) 012012 doi:10.1088/1757-899X/706/1/012012

Table 2. Analyse sample mass.

Samples No.	Sample mass [kg]			
	Geopolymerization 75 [°C]		Autoclaving 160 [°C]	Ripening
	3 [h]	24 [h]	6 [h]	28 [days]
SA1	-	1.880	-	1.878
SA2	-	1.844	-	1.838
SA3	-	1.876	-	1.868
SB1	-	1.848	1.912	1.906
SB2	-	1.850	1.932	1.932
SB3	-	1.840	1.928	1.927
SC1	1.976	-	1.932	1.922
SC2	1.948	-	1.906	1.908
SC3	1.961	-	1.923	1.920

3.2. Compressive strength

Compressive strength tests were performed on standardized cubic samples. Samples were tested after the process of alkaline activation and autoclaving, followed by curing in stable climatic conditions of 50% humidity and 21°C for twenty-eight days. After maturing, tests on the characteristic compressive strength were carried out on a MATEST testing machine. The results of the analysis are presented in Table 3. The value of the characteristic compressive strength does not indicate clear differences between the values. A clear difference occurs in the determined measurement of uncertainty u(x), which indicates that the samples subjected to the autoclave process show a smaller dispersion of compressive strength. This is due to the introduction of the hydrothermal treatment, which significantly affects the course of the activation process, which is confirmed by the morphology presented in the SEM study in section 3.4.

Table 3. Compressive strength after 28 days of ripening.

Samples No.	Compressive strength [MPa]					
	f_c	$f_{cm(3),is}$	$\sigma_{fcm(3),is}$	$u(x)$	$f_{ck,is}$	$f_{ck,is,cube}$
SA1	54.24					
SA2	43.38	49.11	4.452	9.1%	43.11	50.71
SA3	49.70					
SB1	45.53					
SB2	51.36	48.34	2.382	4.9%	42.34	49.81
SB3	48.13					
SC1	48.03					
SC2	47.85	47.93	0.073	0.2%	41.93	49.33
SC3	47.92					

3.3. pH test

One week after the samples were demolded and seasoned in a laboratory room at a humidity of about 50% and a temperature of about 21°C, the samples were immersed to half their height in distilled water. The samples prepared in this way were soaked for a week, adding water to maintain the set level. After a week of soaking, clear eruptions were found on the SA sample. No efflorescence was found on the autoclaved samples.

After soaking, the leachate pH was tested. The results for individual samples are presented in Table 4. It can be stated that the hydrothermal treatment process slightly influenced the reduction of the effluent pH.

Table 4. Examination of the effluent pH after soaking of geopolymer samples.

	SA	SB	SC
pH	11.10	10.63	10.25

3.4. SEM EDS

Examination of the morphology of the samples on a scanning electron microscope (Fig. 2) show clear differences between the sample subjected to the process of only alkaline activation and the samples after alkaline activation and autoclaving. The SA material is characterized by a significant amount of crystallized sodium compounds, which is confirmed by the EDS studies. The EDS test result is shown in Figures 3-5.

Analysis of the content of the elements in the EDS study showed that the Al / Si weight ratio in sample SA is 0.5, while the Na / (Al + Si) weight ratio is 1.0. For the SB and SC samples, the Al / Si weight ratio is 0.6 while the Na / (Al + Si) weight ratio is 0.1. The morphology of the SB sample, relative to the SC sample, shows a greater variation, which results from the process of alkaline activation and autoclaving under various conditions, as described in section 2.3.

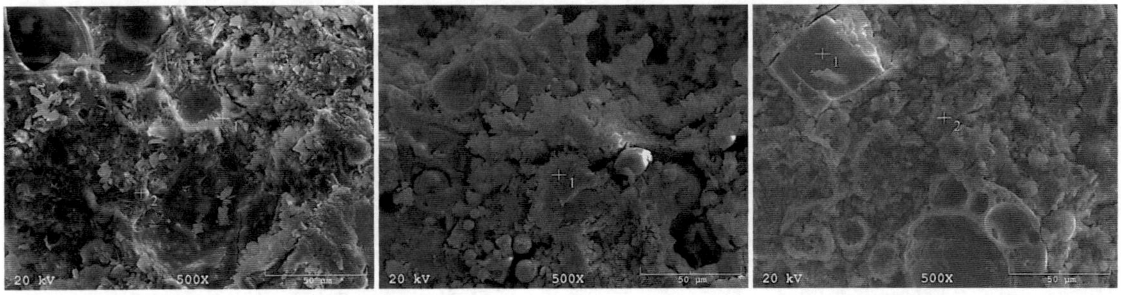

SA SB SC

Figure 2. Morphology of samples after geopolymerization and autoclaving.

IOP Conf. Series: Materials Science and Engineering **706** (2019) 012012 doi:10.1088/1757-899X/706/1/012012

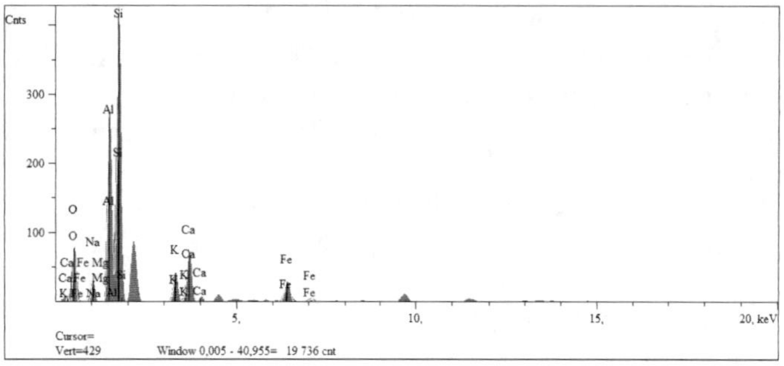

Elt.	Intensity (c/s)	Atomic %	Conc Wt.%	Error 2-sig
O	36.01	46.891	32.405	1.168
Na	15.63	3.517	3.493	0.249
Al	151.95	15.901	18.532	0.331
Si	240.07	26.827	32.545	0.429
K	25.21	1.942	3.281	0.158
Ca	41.81	3.124	5.408	0.191
Fe	21.03	1.797	4.336	0.205
		100.00	100.00	

Figure 3. The characteristic x-ray spectra obtained for the marked points, sample SA.

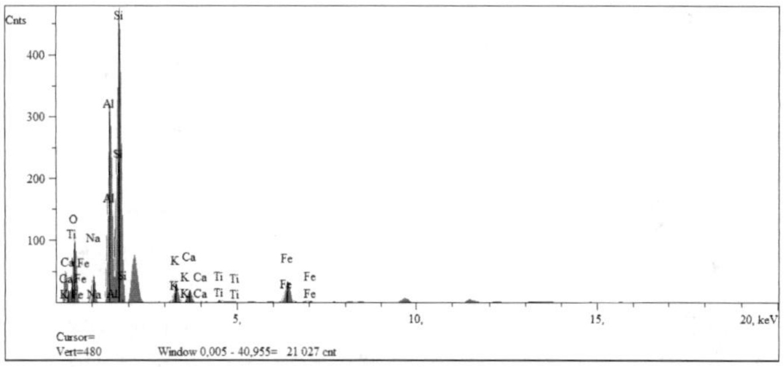

Elt.	Intensity (c/s)	Atomic %	Conc Wt.%	Error 2-sig
O	46.07	47.634	33.491	1.114
Na	20.86	4.151	4.194	0.256
Al	175.67	16.557	19.632	0.328
Si	267.19	27.352	33.759	0.424
K	18.46	1.301	2.236	0.138
Ca	12.80	0.863	1.520	0.122
Ti	3.84	0.246	0.517	0.141
Fe	24.71	1.896	4.652	0.197
		100.00	100.00	

Figure 4. The characteristic x-ray spectra obtained for the marked points, sample SB.

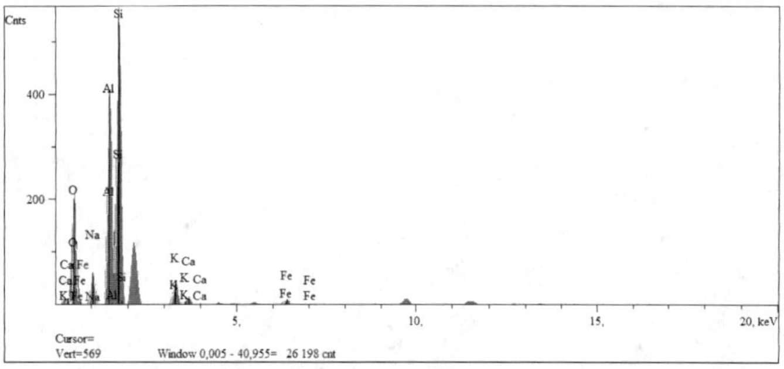

Elt.	Intensity (c/s)	Atomic %	Conc Wt.%	Error 2-sig
O	93.84	57.637	43.833	0.974
Na	28.72	3.911	4.274	0.205
Al	227.57	14.593	18.716	0.267
Si	317.97	21.933	29.280	0.335
K	26.13	1.229	2.284	0.110
Ca	7.04	0.318	0.606	0.082
Fe	7.31	0.379	1.007	0.106
		100.00	100.00	

Figure 5. The characteristic x-ray spectra obtained for the marked points, sample SC.

3.5. Efflorescence on the surface of samples

Samples after the alkaline activation and autoclaving process were placed in containers with distilled water one week after demolding. The water level reached half the sample height. The soaking process took a week. After a week of soaking, a white efflorescence appeared on the surface of the SA sample that evolved on the surface of the sample over the following weeks. Figure 3 shows the observed efflorescence after 28 days from the start of the soaking process.

No white patches were observed on the SB and SC samples.

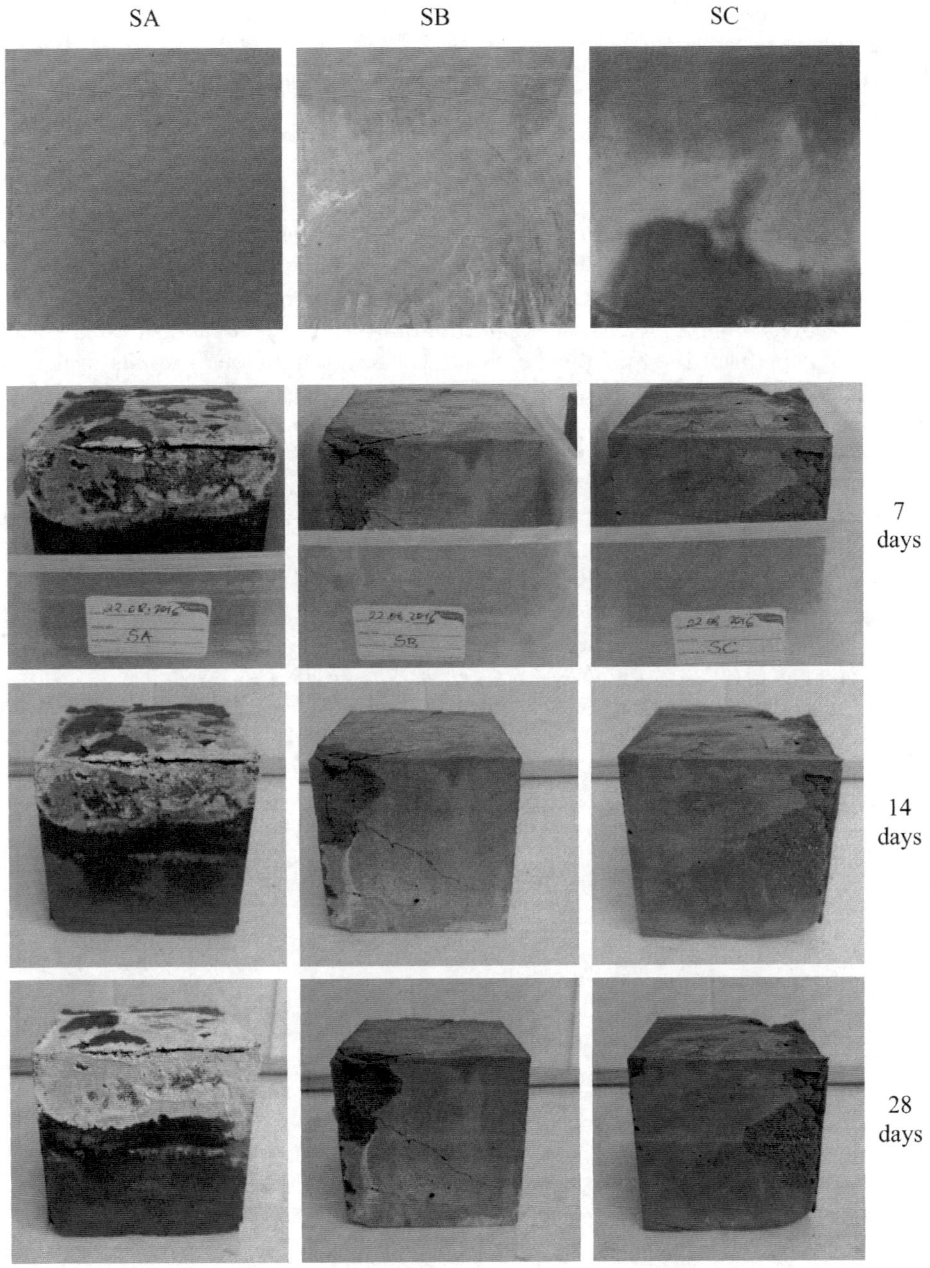

Figure 6. Morphology of samples after geopolymerization and autoclaving surface efflorescence after geopolymersation and autoclaving.

4. Conclusions

Carbonate efflorescence on concrete surfaces, also known as limestone, are formed as a result of carbonation of calcium hydroxide - $Ca(OH)_2$. Calcium hydroxide $Ca(OH)_2$, or portlandite, or colloquially "lime", is one of the hydration products of cement (silicate phases of portland clinker). Carbonate efflorescence arises when calcium hydroxide transported through the capillary pore system gets onto the surface of the concrete element.

IOP Conf. Series: Materials Science and Engineering **706** (2019) 012012 doi:10.1088/1757-899X/706/1/012012

The rate and intensity of carbonation is affected by both the CO_2 concentration and the ambient humidity. In water-saturated concrete, the carbonation process practically does not occur, while in the case of dried concrete, carbonation occurs very slowly. The best conditions for carbonation occur when a layer of water occurs on the capillary surface, in which the rapidly diffusing CO_2 dissolves and reacts with calcium ions. The maximum carbonation rate is in the range of 50% to 70% relative humidity [8].

A similar mechanism occurs in the case of geopolymers, which are often called geopolymer concrete.

The difference in the composition of the mixture of the alkaline activated materials, in relation to traditional concretes, results in the possibility of the appearance on the surface of elements of sodium carbonate heptahydrate, which is revealed quite quickly under appropriate hydrothermal conditions.

One of the solutions to this problem may be the use of hydrothermal treatment using an autoclave.

The impact of superheated steam, together with the increased pressure, results in the increase of capillary channels on the surface of the samples and causes leaching of the excess of unreacted sodium during hydrothermal treatment.

References

[1] Zhang Z, Wang H, Provis J L and Reid A 2013 *Efflorescence: A Critical Challenge for Geopolymer Applications?*, in Concrete Institute of Australia's Biennial National Conference, Toowoomba 2013

[2] Sefflova M and Pavlu T 2016 *Key Engineering Materials* **722** 228-232

[3] Zhang L and Sun L 2017 *Advances in Structural Engineering* **3** (20) 316-330

[4] Shen Q and Pan G 2017 *Materials Science Forum* **902** 74-78

[5] Cui H, Tang W, Liu W, Dong Z and Xing F 2015 *Construction and Building Materials* **93** 522÷527

[6] Duxson P, Provis J L, Luckey G C and van Deventer J S J 2007 *Cement and Concrete Research* **37** (12)1590-97

[7] Karaś S and Miśkiewicz R 2011 *Drogownictwo* **2** 42-48

[8] Omosebiiinni O 2016 *Cement and Concrete Composites* **74** 54-70

IOP Conf. Series: Materials Science and Engineering **706** (2019) 012013 doi:10.1088/1757-899X/706/1/012013

Analysis of the alkali-activated fly ashes from electrostatic precipitators by impedance measurements

J Walter[1], D Mierzwiński[1] and K Kosiba[1]
[1] Institute of Materials Engineering, Faculty of Materials Science and Physics, Cracow University of Technology, al. Jana Pawła II 37, 31-864 Cracow, Poland

E-mail: jwalter@mech.pk.edu.pl

Abstract. Curing the fly ash materials with addition of sand activated by 8 M sodium hydroxide (NaOH) was analysed by impedance spectroscopy. The measurements were carried out for 6 hours from the moment the samples were put into the oven at 75 °C. It was found characteristic similar shape of impedance curves recorded versus time of solidification process. Starting from the curve obtained after 52 min to longer times impedance curve crossed at about 30 Hz. The first lower than 30 Hz frequencies range showed continues decreasing of impedance values with increasing the time. Changes in impedance curve versus time looks different for second frequencies higher than 30 Hz. Temperature of the material increases slightly over temperature of the oven as a result of exothermic reactions which can be seen at impedance vs. time curve as a bulge. The starting of these reactions can be seen more precisely in the second range of the frequency as a continuous increase of impedance value preceded by its initial fast decline. The main purpose of the article is the possibility of using the described method, it can be particularly useful for assessing the starting point of an exothermic reaction and its progress in the case of different masses of heated material.

1. Introduction

Geopolymers are modern materials which can be used in building construction. This term is used to describe a group of materials which can be synthesised from aluminosilicate and high concentrated alkali solution at a temperature range of from ambient temperature to about 100°C [1]. Geopolymers production attracted the interest of scientist because it can be applied parallelly or instead of concrete in many applications.

Another group of materials which can be obtained on a similar way are materials based on fly ash (FA) containing oxides SiO_2, TiO_2, Fe_2O_3, MgO, Na_2O, CaO, K_2O and Al_2O_3 and others [2, 3, 4]. The largest amount of FA is generally a product of burning coal in coal-fired power plants and combined heat and power plants. Burning of coal exceeds temperature of 1300°C. During this process, mineral components that cannot be burned are partly or completely melted. These are usually parts of the rock contained in the coal. The molten particles are largely melted particles pass in the flue gas stream and trapped in the filter as FA. Fly ash can be alkali-activated using high concentrate alkaline in strong water solution. Major advantages of alkali-activated materials are good mechanical performance, low shrinkage with enhanced chemical resistance and thermal stability [5, 6, 7]. Moreover, technology of alkali-activated materials has a reduced environmental impact due to lower CO_2 emission and energy consumption, compare to concretes. FA utilisation is also a significant result of fabrication this kind of materials. Some researchers reported that the electrical conductivity of fly ash based materials depends on the NaOH concentration in activator solution and liquid activator/ash ratio as well as on frequency spectrum [8, 9], others presented dielectrical and electrical study of FA composite [10]. Nowadays impedance spectroscopy (EIS), one of non-destructive methods is more frequently applied for investigations of the materials with capacitive properties [11, 12, 13, 14]. Depending on the capabilities of the device, the tested material, different frequency ranges are used. Our research was carry out to analyse process of curing the FA — sand composite. The frequency range has been chosen that way to be able to carry out a quick measurement because the materials electrical properties are constantly changing while curing. The measurement results were saved as an impedance module in the frequency function.

2. Experimental

The fly ash from Skawina Power Plant (Poland) has been used for production of alcali-activated material. The starting composition consisted by weight 50% of the fly ash and 50% of the building sand. Materials were carefully mixed by 15 minutes using mechanical low speed mixer to obtain evenly distribute particles of powders. The next step was to apply alkaline activator. For this purpose sodium hydroxide (NaOH — Sigma-Aldrich) in concentration of 8 M in distilled water was used. The amount of activator was calculated on the basis of a weight ratio of fly ash to the activator equal to 0.4. Powders with addition of the activator were then mixed by next 15 minutes. Prepared this way material with a slightly liquid consistency was used immediately for filling specially prepared cylindrical mould with two stainless steel electrodes. Vibrator was applied in the next step to remove bubbles of an air from the mould filled with the material.

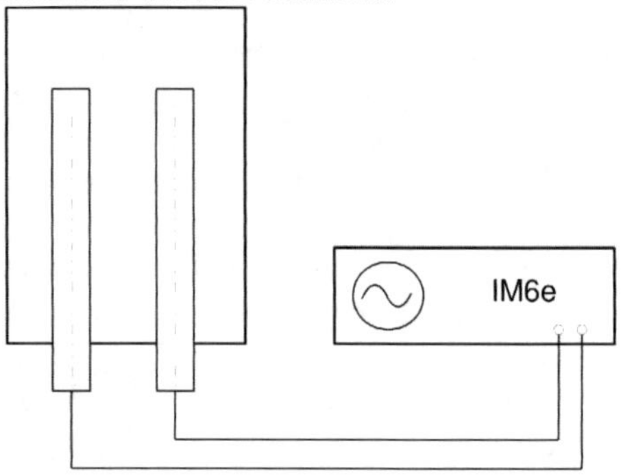

Figure 1. Schema of electrical connections of mould filled with the material consists two electrodes with IM6e workstation.

The mould with total mass equals to 475 g contained 381 g of material was placed in an oven at 75°C. The temperature of the oven was precisely stabilised. Using short electrical cables stainless steel electrodes were connected to the Zahner IM6e workstation to record changing in very low electrical signals generating by the device. The workstation was set in two electrodes mode without any polarisation of the electrodes. The alternating current with amplitude of 50 mV has been applied between electrodes. Frequency of the current were changed during the measurement in the range of 1 Hz to 10 kHz. Each measurement result in impedance curve versus current frequency took 90 seconds. It was taken four measurements per hour for 6 hours during the process of the material curing. The temperature of the material was measured by the calibrated NTC thermistor.

3. Results and discussion

Temperature measurement showed rapid heating of the material after insertion into the oven — Figure 1. The temperature of material reaches the furnace temperature after about 90 minutes and then exceeds it and next slowly decreases keeping oven temperature at the end of recording.

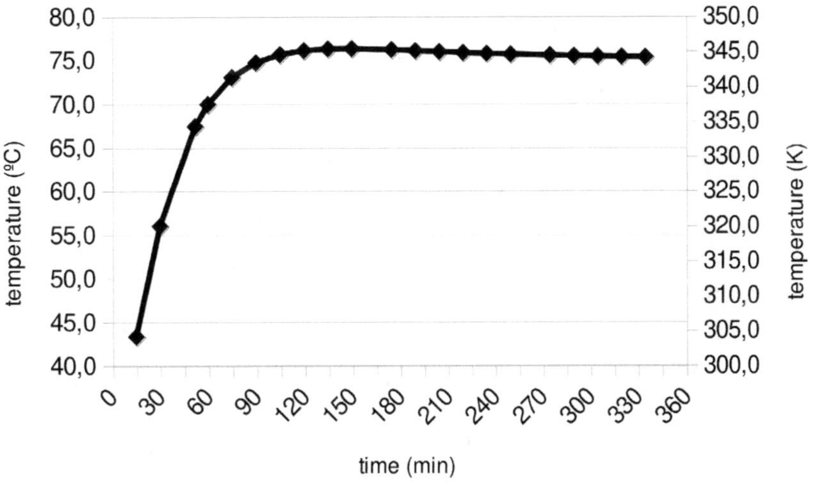

Figure 2. Temperature of material versus time of process.

Figure 3 shows changing of the impedance measured in the range of frequency starting from 1 Hz up to 10 kHz while time of the process increases. All curves crossing themselves at the frequency about 30 Hz. Below this frequency starting from 1 Hz as the time of the solidification process increases, the lower impedance modulus values were always observed in this frequency range. For frequencies higher than 30 Hz the curve shifts are more complicated. The dashed lines showed in Figure 3 describes the curves going down in the graph as the temperature increases. The lowest dashed line (the red one) was recorded after 52 minutes at the temperature equal to 67.4 ºC. This is a result of the material heating up, and a beginning of slow solidification. The measurements taken after more than 52 minutes have shown the shift of the impedance curves towards larger values. The highest temperature of 76.2 ºC was registered after 135 and 150 minutes of the process. A steady decrease in temperature is observed for times longer than 150 min. The temperature of the material is slowly approaching the oven temperature. Figure 4 shows the impedance changes at the frequency of 1 Hz. It can be seen in the temperature range of 67.4 ºC to 76.2 ºC the characteristic course of the curve (Figure 4) corresponding to the increase in temperature caused by the rise of the temperature of heated material and chemical reactions going in it. It was found that the solidification process of a 381 g of the material placed in an oven with a temperature of 75 ºC begins at a temperature of about 65 ºC and the influence of exothermic reactions decreases after about 150 min.

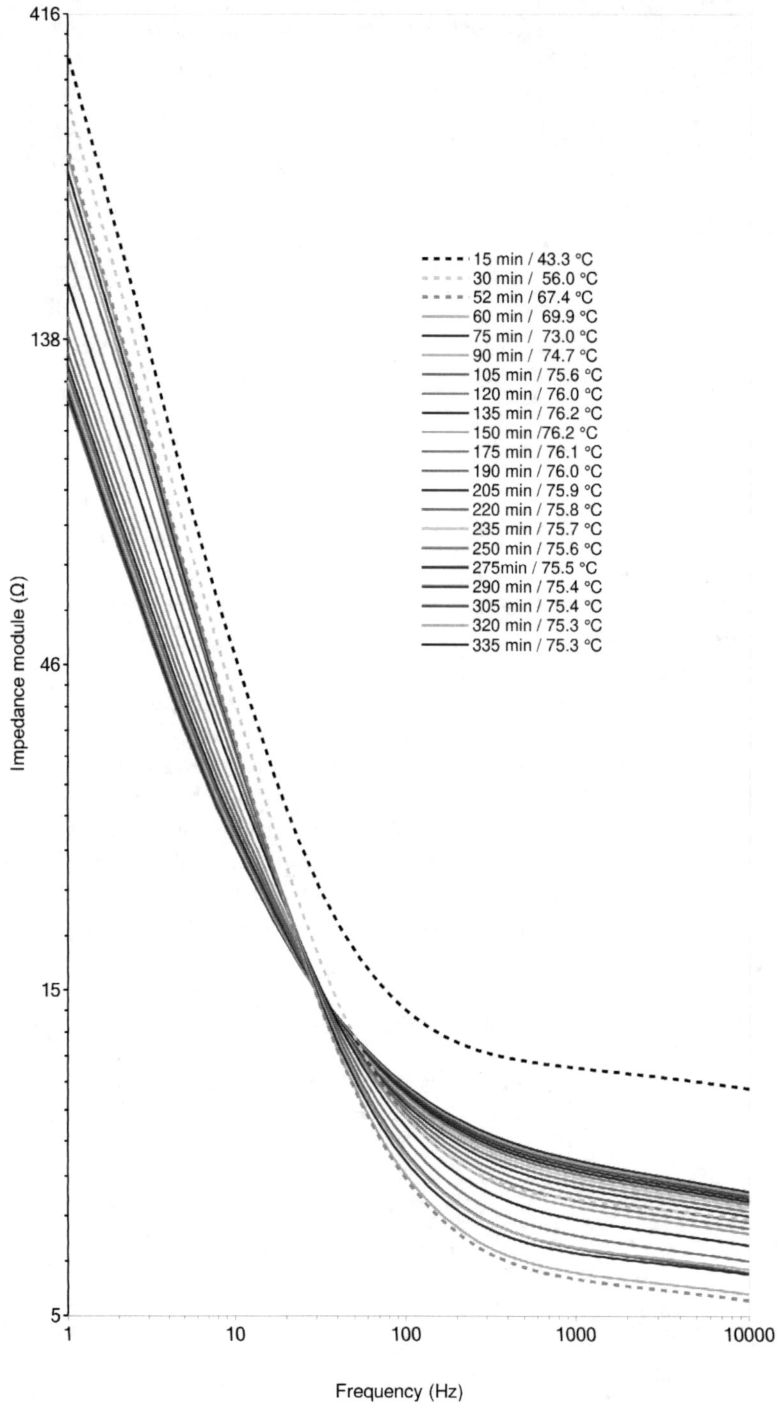

Figure 3. Impedance change versus frequency – time and temperature dependence.

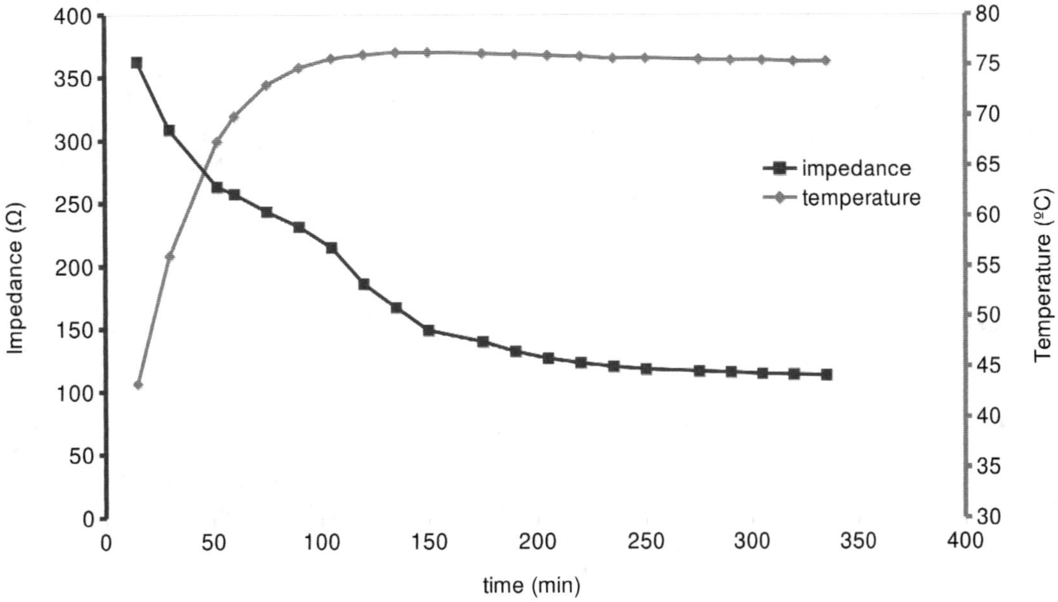

Figure 4. Impedance and temperature changes at 1 Hz versus the time of solidification process.

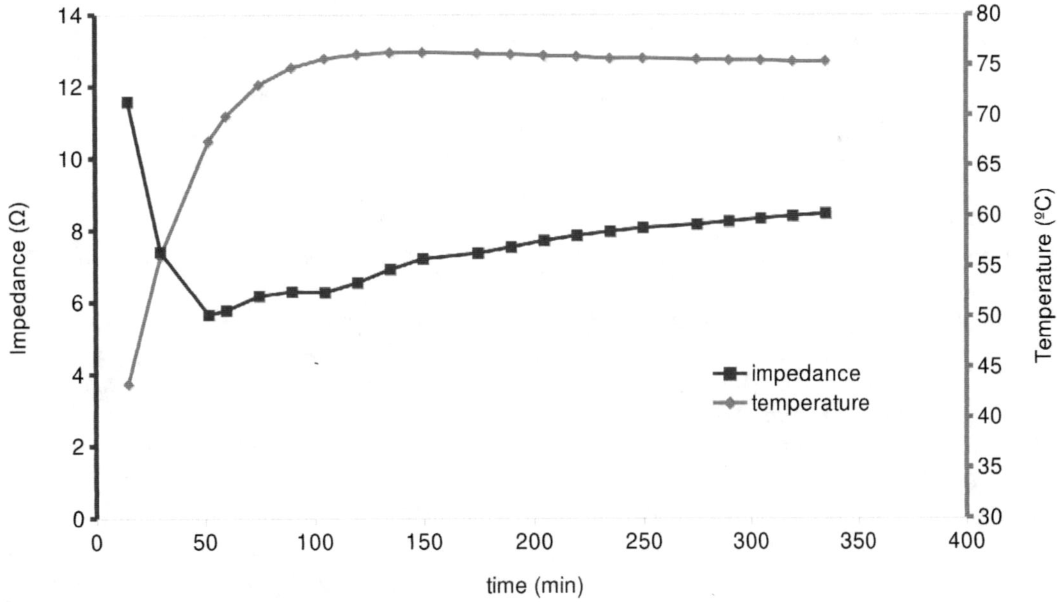

Figure 5. Impedance and temperature changes at 1 kHz versus the time of solidification process.

These processes can be also observed at the higher frequencies. Figure 5 summarises the measurements recorded at the frequency of 1 kHz. Initially the fast decrease in impedance values up to about 52 min. (64.7 °C) was observed with increasing temperature. Next the value of impedance slowly increasing. The same tendency was found for higher frequencies up to 10 kHz as well as for lower than 1kHz but higher than about 30Hz as it was shown in Figure 3. The beginning of exothermic reactions on time axis is easier to recognise in this frequency range because of bigger visible changes on impedance curve.

4. Conclusions

Proposed method of analysing alkali-activated process of the fly ash solidification with addition of sand can be very useful for examining the progress of the process. The measurement which takes 90 seconds is short enough to influence on solidification processes and applied current which is a result of very small voltage amplitude equals to 50 mV cannot also cause any changes in material. Research showed measurable changes in impedance curves versus frequency taken during solidification process of investigated material. The characteristic shape of these curves can be used to check progress of reaction. Heating of mould with semi-liquid material can be observed in lower frequencies as dropping down the impedance values while time increases.

The analysis of impedance curves recorded while process is in progress showed two characteristic ranges of frequencies. The first range from 1 Hz to about 30 Hz (Figure 3) showed a continuous decrease in the impedance value during the process while the changes are getting smaller with increasing frequency. More accurate analysis depicted in Figure 4 for one frequency equals to 1 Hz showed the characteristic curve bulge in the range of material temperature from 64.7°C to about 76 °C. Shape of the curve and its bulge can be useful for estimation of the time of main exothermic solidification processes. The second range beginning from about 30 Hz and ending in the end of recorded frequency seems to be more convenient to find beginning of the exothermic reactions what can be useful especially in case of different mass of heated material.

References

[1] Suwan T and Fan M 2017 *Materials and Manufacturing Processes* **32** 5 461
[2] Saravanan G, Jeyasehar C A, Kandasamy S 2013 *Journal of Engineering Science and Technology Review* **6** 25
[3] Duan W, Lopez M, Juenger M C G, Aughenbaugh K L and Stutzman P 2016 *Front. Mater.* **3** 1
[4] Fernández-Jiménez A and Palomo A 2003 *Fuel* **82** 2259
[5] L. Provis, J.S.J. Van Deventer *Alkali Activated Materials* (Springer, 2014)
[6] Canfield G M, Eichler J, Griffith K *Journal of Materials Science* 2014 **49** (17) 5922
[7] Gunasekara C, Setunge S and Law D W 2017 *ACI Structural Journal,* **114** S60 743
[8] Hanjitsuwan S, Hunpratub S, Thongbai P, Maensiri S, Sata V and Chindaprasirt P 2014 *Cement and Concrete Composites* **45** 9
[9] Hanjitsuwan S, Chindaprasirt P and Pimraksa K 2011 *International Journal of Minerals, Metallurgy, and Materials* **18** 94
[10] Singh P, Bhattacharya B and Singh P K 2017 *Phase Transitions* **90** 3 236
[11] Suryanto B, McCarter W J, Starrs G and Chrisp T M 2017 *Procedia Engineering* **171** 705
[12] Dong B, Qiu Q, Gu Z, Xiang J, Huang C, Fang Y, Xing F and Liu W 2016 *Cement and Concrete Composites* **65** 118
[13] Dong B, Wu Y, Teng X, Zhuang Za , Gu Z, Zhang J, Xing F and Hong S 2019 *Construction and Building Materials* **211** 261
[14] Malkawi A B, Al-Mattarneh H, Achara B E, Mohammed B S and Nuruddin M F 2018 *Construction and Building Materials* **189** 19

IOP Conf. Series: Materials Science and Engineering **706** (2019) 012014 doi:10.1088/1757-899X/706/1/012014

Influence of heat treatment on microstructure and mechanical properties of titanium alloy grade TI6AL4V

R Bogucki[1] and E Hytroś[1]
[1]Institute of Materials Engineering, Faculty of Materials Science and Physics, Cracow University of Technology, al. Jana Pawła II 37, 31-864 Cracow, Poland

E-mail: rbogucki@mech.pk.edu.pl

Abstract: In this work, the effect of heat treatment on the control of the volume share of α phase in a two-phase titanium alloy α + β from the Ti6Al4V grade was analyzed. The applied heat treatments resulted in obtaining a volume share of alpha phase in the range of 60% to 96%. The highest hardness was obtained with a sample share 88% α phase, which was characterized by a microstructure consisting of grains of α phase and β phase regions of the phase precipitates of α'.

1. Introduction

Currently, an increase in demand for engineering materials characterized by high strength combined with low weight has been observed. These criteria is meet light metal alloys, which include titanium or aluminum alloys. Titanium has high strength, plasticity, corrosion resistance and low density. The most popular technical alloy of titanium is an alloy of the Ti6Al4V grade with a two-phase structure α + β, which has applications in various industries such as medical, aviation, automotive, space and sport[1-6].

Alloys with the highest mechanical properties currently include beta or pseudo beta alloys[7]. The chemical composition of these alloys makes it possible to control the volume fraction of the alpha phase by applying appropriate heat treatments. This translates into the ability to control the mechanical properties of these alloys in a wide range [7-9]. The widespread use of titanium alloys is limited by technological problems in its manufacturing and processing.Due to the strong affinity of titanium to oxygen, technological processes must be carried out under inert gas or vacuum. The occurrence of hexagonal crystal latticealso hinders plastic forming and forming of elements with complex shapes [3]. Machining is also difficult due to the low thermal conductivity of titanium [3, 10]. Due to these problems, titanium alloys are used primarily in strategic industries such as aviation, the military where mechanical properties are the most important. In recent years, there has been a significant increase in interest in titanium alloys due to their unique properties. Work is also underway on defect machining techniques that will increase productivity while reducing costs [10-12].

In this work, an attempt was made to assess the impact of volume control on the alpha phase through the use of various heat treatment schemes on the mechanical properties of Ti6Al4V titanium.

2. Experimental procedure

For the research, a two-phase α + β titanium alloy grade Ti6Al4V was used, with the chemical composition shown in Table 1. The phase transition temperatures were determined on a DIL402C

IOP Conf. Series: Materials Science and Engineering **706** (2019) 012014 doi:10.1088/1757-899X/706/1/012014

dilatometer from Netzsch. The beginning of $\alpha \rightarrow \beta$ transformation was determined at 893 °C and the end at 970 °C. Heat treatment was carried out in a Nabertherm P330 tube furnace under an argon atmosphere. In order to obtain a diversified morphology of the α phase, 6 schemes of heat treatment were performed, Table 2.The first three schemes consisted of heating the samples to a temperature of 1050 °C with a heating rate of 10 °C / min and annealing of 15 minutes, which provided a homogeneous β structure. Then the samples were cooled with the oven to the following temperatures: 980 °C, 900 °C, 800 °C, held for 15 minutes, followed by cooling in water (scheme 1, 2 and 3). The next three heat treatment schemes consisted of single-stage heating at temperatures of 980 °C, 900 °C, 800 °C respectively (below the transformation temperature $\alpha \rightarrow \beta$), heating at these temperatures for a period of 15 minutes, followed by cooling in water (scheme 4, 5 and 6). The last heat treatment was aging for 8 hours at 500 °C. The microstructural observations were carried out on a JOEL JSM5510LV scanning microscope with an EDS attachment. Metallographic cuts were made on diamond pastes with a gradation of 6 and 1 μm and etched in Kroll's reagent.

The volume fraction of α phase was determined using the ImageJ program based on electron microscope images with a total analysis of the sample surface of 1 mm^2. The average results are summarized in Table 2. The analysis included only the α phase, ignoring the β and (β + α ') areas. This was dictated by the difficulty in separating individual phases, especially in the area of occurrence of fine dispersion particles of α' phase.

Table 1. Chemical composition of the Ti6Al4V titanium alloy.

Chemical composition of the Ti6Al4V titanium alloy, element content [mass%].							
Al	V	Fe	O	C	N	H	Ti
5,5-6,75	3,5-4,5	<0,30	<0,20	<0,08	<0,03	< 0,008	residue

Hardness measurements were made using the Vickers method, with a 30 kg load on a Hedort hardness tester. Twelve measurements were made for each sample. The average values are shown in Table 2. The tensile test was carried out at room temperature on the MTS CritrtionTM Model 43 testing machine. The elongation and force measurements were recorded using extensometric sensors. Two cylindrical specimens with 5 mm diameter and 25 mm gauge length were prepared.

Table 2. Schemes of heat treatments.

Sample	Type of heat treatment	Temperature domain	Average phase volume fraction α [%]	HV_{30}	YS [MPa]	UTS [MPa]	Elongation A_5 [%]
1s	1050°C/15 min + 980°C /15 min + 500°C/8 h	β+ (α+β)	~87,7	573	1119	1216	16,7
2s	1050°C /15 min + 900°C/15 min+ 500°C/8 h	β+ (α+β)	~91,3	487	1079	1188	6,4
3s	1050°C /15 min + 800°C /15 min+ 500°C/8 h	β+ (α+β)	~96,1	456	1053	1167	4,7
4s	980°C/15 min +	α+β	~60,1	447	1089	1148	16,3

IOP Conf. Series: Materials Science and Engineering **706** (2019) 012014 doi:10.1088/1757-899X/706/1/012014

	500°C/8 h						
5s	900°C/15 min + 800°C /1 min+ 500°C/8 h	α+β	~70,2	363	1058	1123	13,7
6s	800°C/15 min + 500°C/8 h	`α+β	~75,7	337	949	1018	8,1

Results and discussion

The applied heat treatment for titanium alloy grade Ti6Al4V enabled control of the alloy's microstructure and its mechanical properties. The first scheme resulted in obtaining the content of phase α in the range from 88% to 96%. The second scheme obtained α phase content from 60% to 76%. In the first scheme, the heating process was carried out to a temperature of 1050 ° C, which resulted in the transformation of α → β. The next step cooling to temperatures from 980 to 800 ºC, had an aimed of obtaining a varied content of α phase.It was assumed that as the temperature decreases, the content of the α phase should increase. In scheme 2, the annealing process was carried out below the α → β transformation temperature. As in the first scheme, it was assumed that lowering the temperature should lead to an increase in the content of the α phase, which is associated with a decrease in the β phase stability. The additional heat treatment applied in the form of aging at 500 ºC was aimed at increasing the content of α 'phase precipitates in the β phase. Observations of the microstructure revealed a diverse morphology of the alloy structure, Fig. 1.

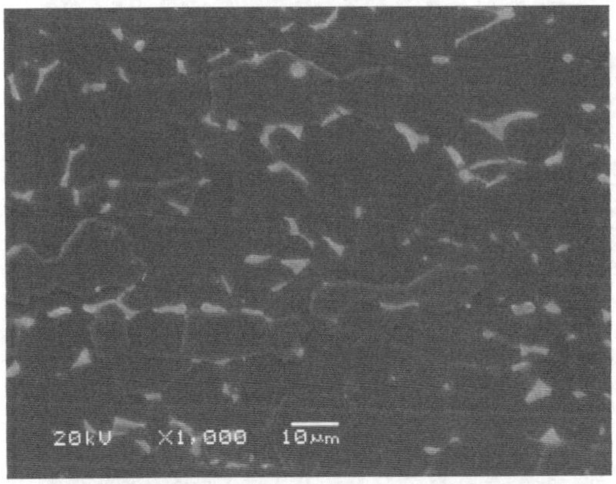

a)

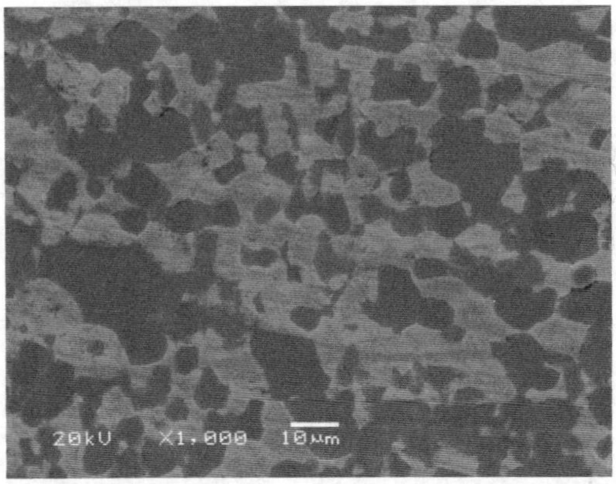

d)

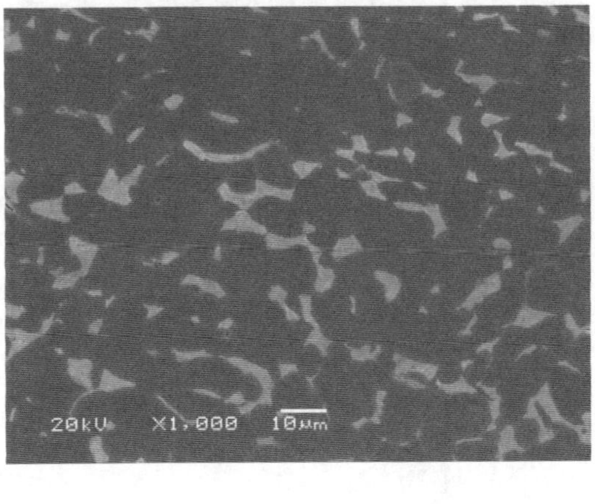

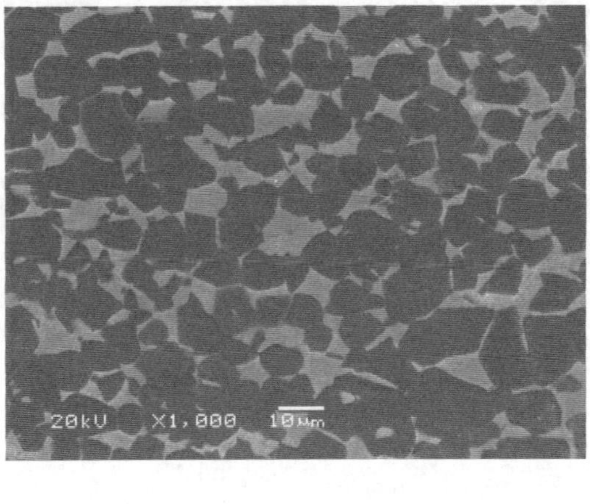

IOP Conf. Series: Materials Science and Engineering **706** (2019) 012014 doi:10.1088/1757-899X/706/1/012014

b)

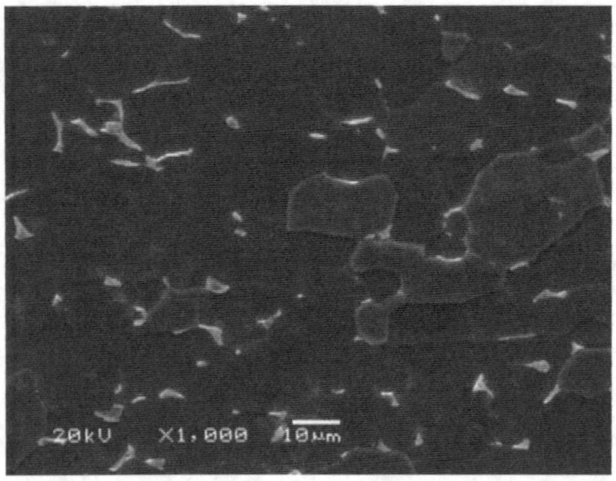

c)

e)

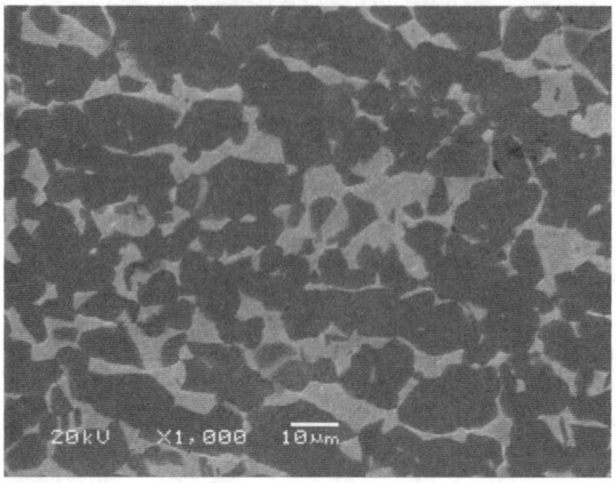

f)

Figure1. Microstructure of Ti6Al4V alloy after heat treatment: a) scheme 1s, b) scheme 2s, c) scheme 3s, d) scheme 4s, e) scheme 5s, f) scheme 6s.

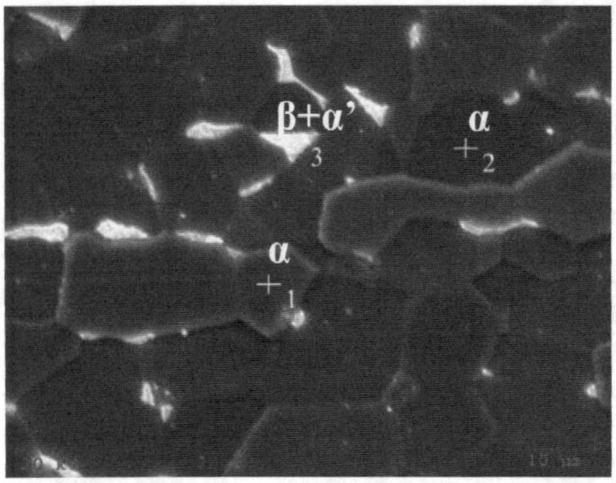

a)

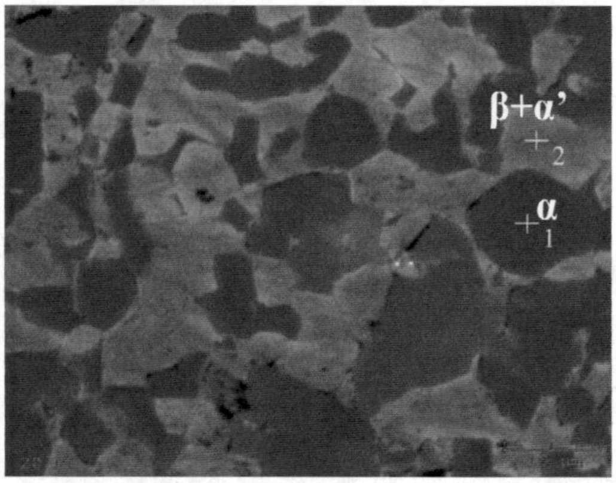

b)

Point 1a

Elt.	Line	Intensity (c/s)	Error 2-sig	Atomic %	Conc Wt. %
Al	Ka	62,10	1,582	15,454	9,331
Ti	Ka	733,11	4,922	83,838	89,862
V	Ka	6,04	1,735	0,708	0,808

Point 1b

Elt.	Line	Intensity (c/s)	Error 2-sig	Atomic %	Conc Wt. %
Al	Ka	66,29	1,647	15,866	9,599
Ti	Ka	758,46	5,019	83,585	89,774
V	Ka	4,86	1,790	0,549	0,627

Point 2a

Elt.	Line	Intensity (c/s)	Error 2-sig	Atomic %	Conc Wt. %
Al	Ka	57,53	1,551	14,443	8,678
Ti	Ka	736,16	4,943	84,533	90,161
V	Ka	8,70	1,815	1,024	1,161

Point 2b

Elt.	Line	Intensity (c/s)	Error 2-sig	Atomic %	Conc Wt. %
Al	Ka	53,80	1,511	12,976	7,727
Ti	Ka	758,88	5,042	82,912	87,650
V	Ka	36,70	2,053	4,112	4,623

Point 3a

Elt.	Line	Intensity (c/s)	Error 2-sig	Atomic %	Conc Wt. %
Al	Ka	53,83	1,508	13,615	7,114
Ti	Ka	690,71	4,811	79,076	84,662
V	Ka	62,24	2,218	7,309	8,224

Figure 2. Microstructure and results of the microanalysis of chemical composition in micro areas: a) sample 1s, b) sample 2s.

In order to confirm the occurrence of α and β phases, a microanalysis of the chemical composition was carried out for selected samples using Energy-dispersive X-ray spectroscopy, Figure 2. The difference in vanadium and aluminum content in the alpha and beta phases was observed. The β phase stabilizing element is vanadium in the analyzed alloy. Areas richer in this element were a mixture of the β phase with the precipitations of the α' phase which was formed during the aging process. The α phase were areas rich in aluminum, which is the basic element stabilizing this phase. The highest level of hardness and strength combined with high toughness was obtained for the lowest percentage of phase α (samples 1s and 4s), table 2. This is probably related to the higher proportion of phase β which is characterized by higher mechanical properties due to the body-centered cubic crystal structure. The increase in the volume fraction of α phase with hexagonal close-packed crystal structure led to a decrease in strength properties as well as to a reduction in alloy plasticity. The highest mechanical properties were obtained for the first heat treatment scheme and the content of α phase at 88%. The hardness obtained was 573 HV and the tensile strength and yield strength 1216 MPa and 1119 MPa, respectively, with a relative elongation A5 of 16.7%. Similarly, for the second scheme, the highest strength parameters were achieved for the lowest volume fraction of phase α of 60.1%. The hardness was 447 HV, the tensile strength and yield strength were 1148 MPa and 1086 MPa at 16.3% elongation. Despite the lower content of α phase in the second scheme, lower mechanical properties were obtained. The reasons should be seen in the method of heat treatment. Probably the use of heat treatment with the α → β transformation led to better distribution of the α' phase precipitates after the aging process. Confirmation of this, however, requires research using transmission microscopy.

Conclusions

1. The analyzed titanium alloy Ti6Al4V has a two-phase α + β microstructure, where the volume fraction of the alpha phase and its morphology depends on the heat treatment applied.
2. After applying additional aging at 500 °C, a fine dispersion of α′ is precipitated in the supersaturated β phase.
3. The best combination of mechanical properties was obtained after heat treatment according to scheme 1s. The alloy had a yield strength of 1119 MPa, a tensile strength of 1216 MPa, and an elongation of 16.4%.
4. The microstructure of the alloy after heat treatment according to the 1s scheme was composed of grains of the α phase and the β phase in which the fine-dispersion α' phase nucleated.

Reference

[1] Terlinde G T, Duerig T W and Williams J C 1983*Metall. Mater. Trans.* A **14** 2101-15
[2] Duerig T W, Albrecht J, Richer D and Fischer P 1982 *Acta Metall.* **30** 2161-72
[3] Oczkoś K E 2008*Mechanik* **8-9**
[4] Dudek Ł, Hryniewicz T and Rokosz K 2016 *Autobusy : technika, eksploatacja, systemy transportowe* **17** (8) 62-66
[5] Titanium Alloy Guide, RMI Titanium Company, Jan/2000 (2)

[6] Boyer R R and Brigs R D 2005 *J. Mater. Eng. Perform.* **14** (6) 681-685
[7] Bogucki R, Mosór K and Nykiel M 2014 *Arch. Metal. Mater.* **59** 1269-73
[8] Dąbrowski R 2011 *Arch. Metal. Mater.* **56** 217-221
[9] Dąbrowski R 2011 *Arch. Metal. Mater.* **56** 703-707
[10] Zębala W, Gawlik J, Matras A, Struzikiewicz G and Ślusarczyk Ł 2014 *Precision Machining* **581** 409-414
[11] Bojko Ł, Ryniewicz A M, Bogucki R and Pałka P 2015 *Prz. Elektrotech.* **91** (5) 29-32
[12] Żyra A, Bogucki R and Skoczypiec S 2019 *Arch. Metal. Mater.* **64** (3) 1005-10

IOP Conf. Series: Materials Science and Engineering **706** (2019) 012015 doi:10.1088/1757-899X/706/1/012015

Development of a stabilized natural fiber-reinforced earth composite for construction applications using 3D printing

G Silva[1], L Quispe[1], S Kim[1], J Nakamatsu[2] and R Aguilar[1]

[1]Engineering Department, Pontificia Universidad Católica del Perú PUCP, Av. Universitaria 1801, Lima 32, Lima, Peru
[2]Science Department, Pontificia Universidad Católica del Perú PUCP, Av. Universitaria 1801, Lima 32, Lima, Peru

Email: raguilar@pucp.pe

Abstract. The application of additive manufacturing technologies to the construction industry has a wide range of advantages from the economic, social and design flexibility point of view. However, most of up to date research studies have been performed using ordinary Portland cement (OPC)-based mortars and concretes. Therefore, the objective of this article is to explore the development of an eco-friendly earth composite reinforced with natural sisal fibers and chemically stabilized with a hydraulic binder. Analysis of the workability by shear vane tests was performed on fresh earth samples to find the optimum water content in sisal fiber-reinforced earth stabilized with OPC. Afterward, the effect of the addition of OPC on the hardening process was evaluated through shear vane and Vicat needle tests from 0 to 180 min after mixing. The results indicate that water content, fiber addition and OPC replacement have a strong influence on the shear yield strength. Furthermore, the presence of chemical stabilizers as OPC accelerates the hardening process allowing a faster layer-by-layer deposition. This low cost and eco-friendly preliminary earth-based composite can be used for 3D printing applications in the construction industry.

1. Introduction

3D printing was defined by the American Society of Testing Materials (ASTM) as the manufacture of objects by depositing material using a print head [1]. The literature indicates that materials used for 3D printing applications for construction should be designed according to the particular characteristics of the printing system [2] and should ensure flow ability with a balance between a self-compacting mixture and a non-slump mixture [3]. During the printing process, this material must satisfy specific conditions of extrusion, shape retention, capacity to control deformation to withstand successive layers and good adhesion [4–6].

From a material, economic and environmental point of view, earth has proven to be a suitable alternative building material that properly used can provide safe solutions that are also thermally and acoustically efficient [7–9]. However, compared to other traditional building materials (e.g. concrete or fired clay), untreated raw earth has several limitations and disadvantages. Some of its negative points are its low resistance to compression and tensile loads, vulnerability to water erosion, low resistance to dynamic loads and high cracking during the drying process [7 – 10]. Chemical stabilizers and fiber reinforcement have been actively investigatedfor modifying this material obtaining earth-based composites with improved compressive, tensile and flexural strength [11] and enhanced durability [12]. Aerial binders such as quick lime and slaked lime [13], hydraulic binders such as OPC

[14], geopolymers such as fly-ash-based geopolymers [15] and pozzolana-based geopolymers [16] and biopolymers such as alginate [11], carrageenan [17] and chitosan [18] have been used as chemical stabilizers of earth-based construction materials. In the case of natural fiber reinforcement, a wide range of natural plant and animal fibers were employed in earthen buildings materials [9, 11, 19].

3D printing projects for construction such as Contour Crafting (CC) [20], the 3D concrete printing project of Loughborough University [21], D-shape [22]and the proposal for large-scale 3D printing of high-performance concrete developed by Gosselin et al. [23]stand out in this area. All of these projects have been designed with OPC-based mortars or concretes. There are limited applications related to the use of earth as a building material for 3D printing applications. Perrot et al. [24] used stabilized earth with alginate biopolymer for the generation of an extrudable material. The use of this agent significantly increased the gain rate of the yield strength allowing a stable layer-by-layer deposition. On the other hand, the Italian company Wasp3D presented in 2018 the first printed house built with stabilized earth and reinforced with rice residues [25]. The material used to build one-story modular houses was composed of 25% earth (30% clay, 40% silt and 30% sand), 65% rice residues and 10% lime. Therefore, the objective of the present research is to explore the development of a new earth-based composite composed of earth, natural sisal fibers and OPC for 3D printing applications. A detailed description of the followed methodology based on the analysis of the workability and the hardening process to obtain an earth-based composite adequate for 3D printing applications is presented.

2. Materials and experimental plan

2.1. Description of raw materials

Earth used in this study was soil extracted from Callao (Peru). As shown in Figure 1a, this raw material was inadequately for 3D printing applications due to the high content of gravel and lumps. Therefore, the material was passed through ASTM sieve No.20 to remove particles larger than 0.875mm obtaining a ratio of the maximum particle size to the diameter of the delivery system pipe of 1/29 (less than the recommended maximum ratio of 1/10).This process allowed obtaining the sieved earth shown in Figure 1b that exhibited a particle size distribution envelope given in Figure 1c. The granulometric analysis of the earth was performed following the statements of UNE-EN ISO 17892-4:2019 [26] and indicated a d90= 250 μm, d50=25 μm and d20=1.8 μm. Approximate average contents of coarse sand, fine sand, silt and clay of 15%, 41%, 21% and 23%, respectively, were defined and demonstrated that this material fulfilled the recommendation of a maximum content of gravel and sand of 80% given by Lecompte& Perrot [27].

In addition, the sieved earth was subject to a physical characterization campaign following the standard UNE-EN ISO 17892-12:2019 [28] (see Table 1). The earth presented a liquid limit of 29.6%, a plastic limit of 19.9%, and a plasticity index of 9.7. Based on these results, the earth was classified as inorganic clay soil. Finally, the moisture content and apparent specific gravity of the raw material were measured at 3.4% and 2.6, respectively.

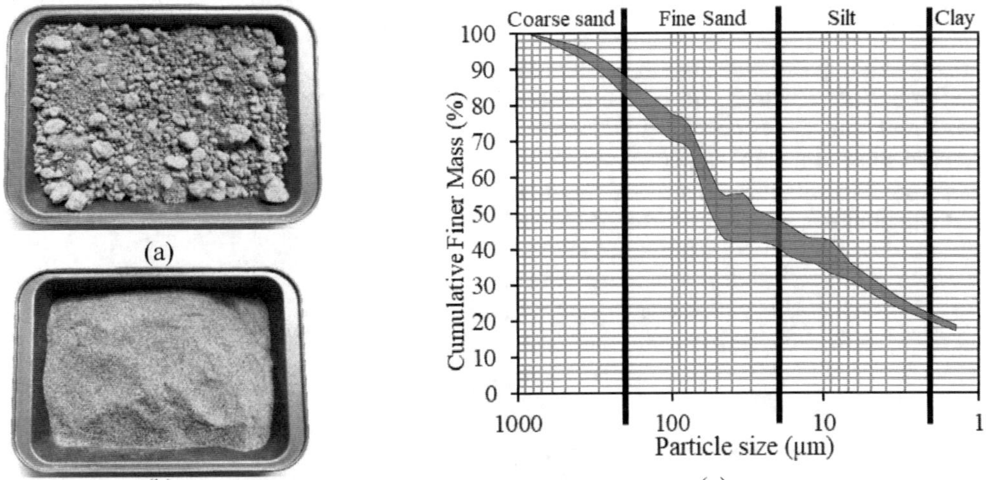

(b) (c)

Figure 1. Raw material: (a) Raw earth; (b) Sieved earth; (c) Particle size distribution envelope of sieved earth.

Table 1. Physical properties of sieved earth.

Liquid limit (%)	Plastic limit (%)	Plasticity index	Moisture content (%)	Specific gravity
29.6	19.9	9.7	3.4	2.6

Natural sisal fibers were used as short-random reinforcement of the earth-based material. Sisal fibers present a density of about 1.30-1.45 g/cm^3 [29] and are composed of around 58-63% cellulose, 20-24% hemicellulose, and 7-9% lignin by weight[30]. In the present research, sisal fibers were obtained from commercial products (ropes) and cut to a length of approximately 10 mm. The average diameter of sisal fibers was calculated around 137 μm (aspect ratio l/d of 73), measured with a Mitutoyo dial thickness gauge (model 7301).The tensile strength of sisal fibers was investigated by direct tensile tests carried out with an MTS Exceed 42.053 machine with 5kN at a displacement rate of 6 mm/min. Sisal fibers showed a tensile strength, E-modulus and elongation capacity of 508 MPa, 27 GPa and 2%, respectively.

OPC is a fine-grained material produced by gypsum and clinker inter grinding. Clinker, the major component of cement (95%), is a pyro processed hydraulic material mainly composed by tricalcium silicate (C_3S), dicalcium silicate (C_2S), tricalcium aluminate (C_3A) and tetracalcium aluminate (C_4AF), where both calcium silicates compounds are the most important for earth stabilization[14]. For this research, Portland Cement SOL type I was used as a chemical stabilizer of the fiber-reinforced earth-based composite.

2.2. Characterization of the fresh properties on earth-based composites

Trial mixes of earth-based composites were subjected to shear vane tests to evaluate their suitability for 3D printing applications. As Le et al. [31] stated, shear vane tests can provide more valuable and scientific information regarding workability than conventional methods as slump, compacting factor and flow tests. A Humboldt H-4227 field shear vane set with vanes with a diameter: height ratio of 2, as specified by ASTM D2573/D2573M−18 [32], was employed to perform the workability assessments of all earth-based composite mixtures. Due to the relatively low shear stresses presented by the material immediately after mixing, a 25.4 x50.8 mm vane (conversion factor of 0.5) was used during workability analysis. All shear yield strength measurements were performed immediately after mixing in a container with the dimensions shown in Figure 2a. Large dimensions of the container allow taking nine measurements per test avoiding boundary effects as shown inFigure 2b.

An evaluation of the shear yield strength gain overtime was performed to study the effect of the OPC on the hardening process of the earth-based composites. The tests were performed at 0, 30, 60, 120 and 180 minutes after mixing. In this stage, in addition to 25.4 x 50.8 mm vane, 20x40 mm and 16x32 mm vanes (conversion factor of 1 and 2, respectively) were used to study the fiber-reinforced earth-based composites stabilized with OPC due to the high shear stresses registered from 30 minutes. The hardening evaluation was complemented with Vicat needle tests (Figure 2c). Measurements were taken from 0 to 180 minutes after mixing in intervals of 10 min in a sample poured in a truncated cone mold with a dimension of 70x60x40 (base diameter x top diameter x height).

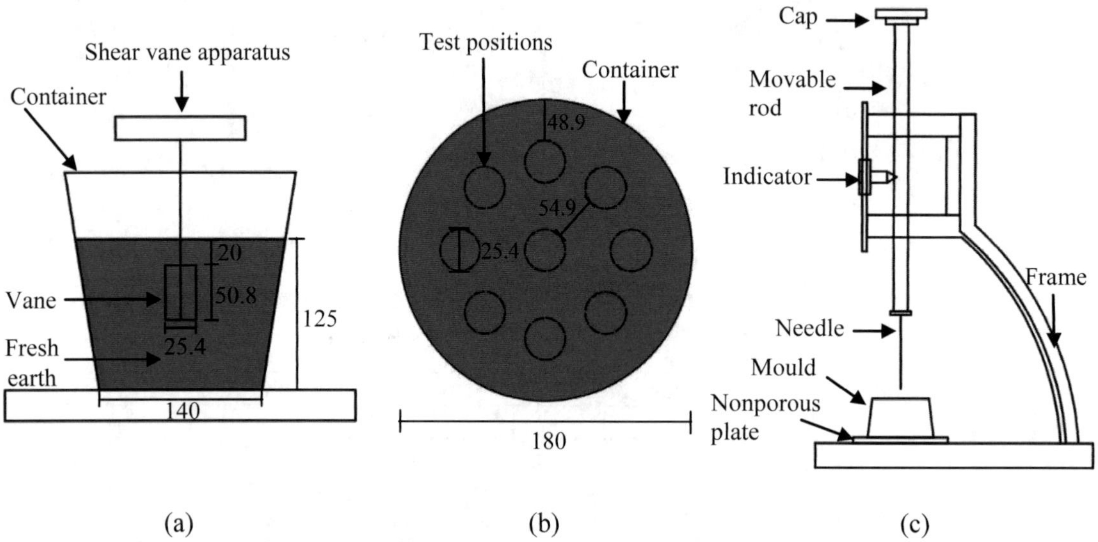

Figure 2. Experimental procedure diagrams: (a) Side view of shear vane tests; (b) Plan view of shear vane tests with measuring points; (c) Vicat needle tests.

2.3. Preparation of samples

Earth-based matrix (EM), fiber-reinforced earth-based matrix (FEM) and fiber-reinforced earth-based matrix stabilized with OPC (CFEM) were prepared to study their effect on the workability and hardening process of earth-based composites. Earth, sisal fibers and OPC were dry-mixed to ensure homogenous distribution. After that, tap water was gradually added and mixed until a homogenous paste was obtained. All mixing process was executed using a mortar mixing machine UTG-0130 (UTEST) with a capacity of 10 L. Once pouring the fresh earth-based composite in the container for shear vane tests and truncated cone mold for Vicat needle tests, samples were kept under room temperature and relative humidity of 18 °C and 74 %, respectively.

Different contents of water were studied for EM, FEM and CFEM (a summary of all trial mixtures is presented in Table 2). The water content is expressed in terms of the weight ratio (wt.%) of water to the total composite (earth + water + OPC). Water content was varied for the three earth-based composites to reach suitable workability in terms of the shear yield strength for 3D printing applications. For this research, a maximum shear yield stress of 2.2 kPa was defined based on the recommendations given by Perrot et al. [24] for an extrusion machine with a maximum working pressure of 20 bar. Fiber addition and OPC replacementat1% (wt. %) of fibers to the total solids (earth + OPC) and 13 % (wt. %)of OPC to total solids, respectively, were used for the reinforcement and chemical stabilization of EM. OPC replacement value was chosen on the basis of the recommendation given by UNE 41410:2008 [33] that stated a maximum of 15% of stabilizers in earth-based materials.

Table 2. Summary of mixture proportions used for the preparation of earth-based composites.

Earth-based composites	N°	Water content (wt. % of the total composite)	Fiber addition (wt. % of solids)	OPC replacement (wt.% of solids)
Earth-based matrix (EM)	1	18.0	0	0
	2	19.0		
	3	20.0		
Fiber-reinforced earth-based matrix(FEM)	4	18.0	1	0
	5	19.0		
	6	20.0		
	7	21.0		
	8	21.5		
	9	22.0		
Fiber-reinforced earth-based matrix stabilized with OPC (CFEM)	11	20.6	1	13
	12	21.5		
	13	22.5		
	14	24.3		

3. Results and discussion

3.1. Evaluation of the workability of earth-based composites

Stress growth experiments were performed using a shear vane apparatus to obtain the shear yield strength of all earth-based composites immediately after mixing. During stress growth shear tests, the material is subjected to a constant strain rate at low rotational speeds (Figure 3a). It is important to execute the tests at low speeds since the conversion from maximum torque to yield strength is valid only for low rpm. The conversion from torque to yield strength is obtained with equation (1) for a vane with H/D=2 [32]:

$$\tau_y = \frac{6T}{7\pi D^3} \times 10^6 \tag{1}$$

Where T is the maximum torque, τ_y is the yield strength while H and D are the height and diameter of the vane, respectively. The conversion factors are expressed with respect to a standard vane of 20 x 40 mm.

The results of shear vane tests on earth-based composites are shown in Figure 3b. The results indicate that the water content has an indirect exponential relationship with the shear yield strength for all the three earth-based composites. In the case of EM, the water content necessary to obtain a yield strength of less than 2.2 kPa was 20%. When sisal fibers were added, for constant water content, the shear yield strength significantly increased since sisal fibers have a water absorption capacity of 125% and also high surface area due to its high aspect ratio (L/D = 74). Therefore, a water content of 21.5% in FEM was used to obtain a shear yield strength of 1.5 kPa. Finally, an OPC replacement at 13% (wt. %) again affect severely the shear yield strength of the material, an increment of 380% was observed for a water content of 21.5%. This increment of the yield strength caused by the replacement of earth by OPC can be explained by the significantly lower specific surface area of OPC by its finer particle size (usually from 0.5 to 50 um) that increases the water demand. Consequently, the water content necessary to obtain a CFEM for printing applications using an extrusion machine of 20 bar was 24.3% (wt. %).

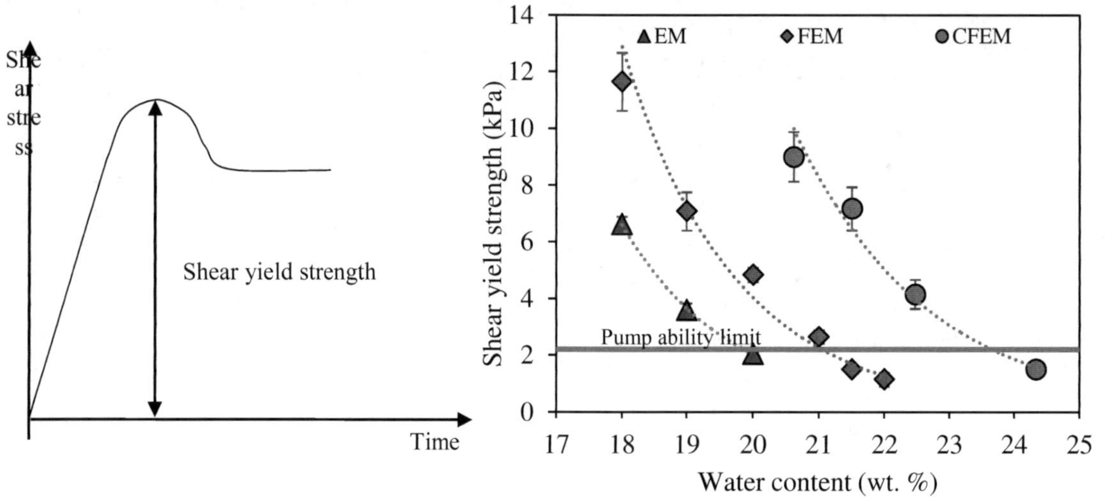

(a) (b)

Figure 3. Shear vane tests: (a) Scheme of the stress grow tests; (b) Influence of the water content in the shear yield strength for EM, FEM and CFEM.

3.2. Evaluation of the hardening process of earth-based composites

The hardening process was evaluated on FEM with a water content of 21.5% and CFEM with a water content of 24.3% through shear vane tests at 0, 30, 60, 120 and 180 min and the results are displayed in Figure 4a. The results indicate the great influence of the hydraulic binder presence on the hardening process. After 60 minutes, the stabilized earth-based composite (CFEM) reached a yield strength of 29 kPa. Wangler et al. [34] stated that the vertical compression load applied to a given layer by a printed structure is equal to ρgH where ρ is the density in the fresh state of the printed material (1730 kg/m^3 for this research), g is the gravity constant and H is the height of the printed structure above a given layer. Therefore, the developed CFEM is able to support the weight of a 1 m high wall after 1 hour. After 120 min, the stabilized earth-based composite is able to sustain a 3 m high wall. However, this rapid gained of yield strength can be undesirable if the material cannot be extruded in a few minutes, a concept defined as open time or printability window. On the other hand, FEM did not exhibit major evolutions on its shear yield strength. This can be explained due to the hardening of earth-based materials without chemical stabilizers is based on free water loss, which is a very slow process, especially at high relative humidity environments (e.g. 70%). On the contrary, hydraulic compounds of OPC as C_3S and C_2S can set and harden in contact with water and produce calcium hydroxide crystals responsible for cation exchange and flocculation and agglomeration, and calcium silicate hydrate (C-S-H) that bring strength to the earth matrix [14]. In addition, ettringite, which is the first hydrate product to crystalize (1 hour after mixing), contributes to stiffening and setting during early ages[35]. The results of shear vane tests are compatible with Vicat needle tests, a conventional method to evaluate the setting time of OPC pastes. As in the case of shear vane tests, the earth-based material with no OPC replacement (FEM) did not exhibit stiffening up to 180 min. However, the addition of OPC at 13% (CFEM) alter the penetration resistance to the Vicat needle of the earth-based composite. Furthermore, the data of the latter are well fitted to a hyperbolic secant curve (see Figure 4b). This trend is in accordance with Knudsen [36]who proposed linear and exponential hyperbolic models for the hydration of cement-based materials.

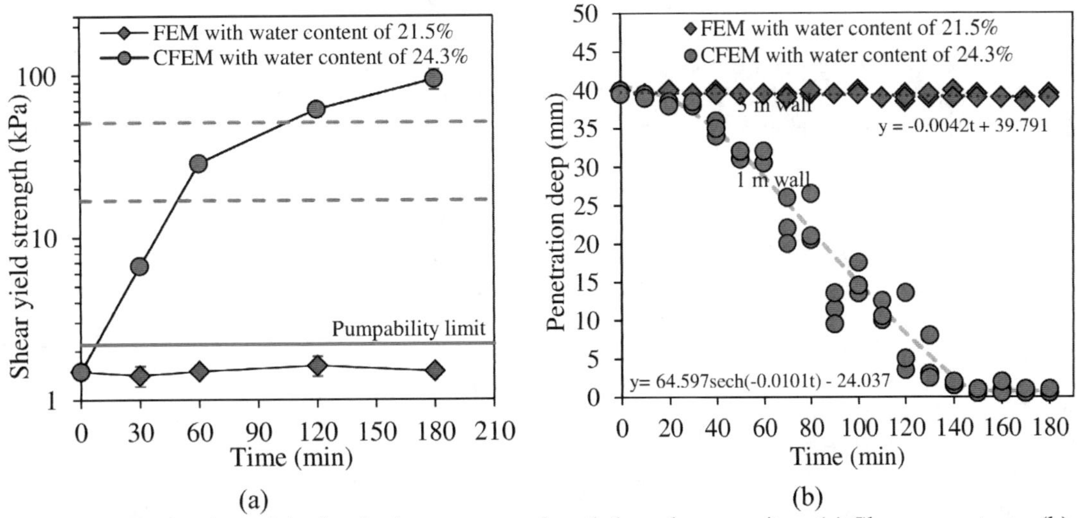

Figure 4. Evaluation of the hardening process of earth-based composites: (a) Shear vane tests; (b) Vicat needle tests.

4. Conclusions

This article was aimed to analyze the workability and hardening process of earth-based composites in the context of 3D printing for construction applications. An earth-based material with a water content of 24.3% (wt. %) reinforced with 1% (wt. %) of sisal fiber sand chemically stabilized with 13% (wt. %) of OPC was developed. This building material has a shear yield strength immediately after mixing of 1.5 kPa, making it suitable for 3D printing applications using an extruder with a minimum working pressure of 20 bar. Furthermore, the analysis of the hardening process allowed to demonstrate that a layer of this new earth-based composite can sustain a 3 m wall after 120 min. This enhanced hardening of the earth-based material is fundamental to achieve high-speed construction processes.

Acknowledgments

This research was supported by Consejo Nacional de Ciencia y Tecnología (CONCYTEC) and Servicio Nacional de Capacitación para la Industria de la Construcción (SENCICO) of Peruunder the Contract N° 130-2018-FONDECYT.The authors acknowledge Compañía Minera Agregados Calcáreos S.A (COMACSA) for facilitating the use of its equipment. The authors also thank the support of Rossemary Enciso, Kevin Huamaní and Diana Zavaleta during the characterization of the raw materials. Guido Silva acknowledges CONCYTEC for funding his Ph.D. studies under the scholarship N° 10-2018-FONDECYT/BM.

References

[1] ASTM International ISO/ASTM 52900 2015
[2] Buswell R A, Thorpe A, Soar R C and Gibb A G 2008 *Automat Constr* **17** (8) 923-929
[3] Panda B and Tan M J 2018 *Ceram Int* **44** (9) 10258-65
[4] T S Rushing, Stynoski P B, Barna L A, Al-Chaar G K, Burroughs J F, Shannon J D and M P Case 2019 In 3D Concrete Printing Technology137-160
[5] Kazemian A,Yuan X, Cochran E and Khoshnevis B 2017 *Constr Build Mater* **145** 639-647
[6] Paul S C, Tay Y W D, Panda B and Tan M J 2018 *Arch Civ Mech Eng* **18** (1) 311-319
[7] Aymerich F, Fenu L and Meloni P 2012*Constr Build Mater* **27** (1) 66-72
[8] Pacheco-Torgal F and Jalali S 2012*Constr Build Mater* **29** 512-519
[9] Sharma V, Vinayak H K and Marwaha B M 2015 *Constr Build Mater* **93** 943-949
[10] Miccoli L, Müller U and Fontana P 2014*Constr Build Mater* **61** 327-339
[11] Galán-Marín C, Rivera-Gómez C and Petric J 2010 *Constr Build Mater* **24** (8) 1462-68

[12] Donayre A, Sanchez L F, Kim S, Aguilar R and Nakamatsu J 2018 *IOP ConfSer-Mat SciEng* **416** (1) 012044
[13] Oti J E, Kinuthia J M and Bai J 2009 *P I Civil Eng-Eng Su* **162** (4) 229-237
[14] Prusinski J R and Bhattacharja S 1999 *TransportRes Rec* **1652** (1) 215-227
[15] Cristelo N, Glendinning S, Miranda T, Oliveira D and Silva R 2012 C *Constr Build Mater* **36**727-735
[16] Alvarez S Y 2019 Comparación de las propiedades mecánicas de unidades y prismas de bloques de tierra comprimida estabilizada con cemento y geopolímero de puzolana *Master thesis in Civil Engineering PUCP*
[17] Nakamatsu J, Kim S, Ayarza J, Ramírez E, Elgegren M and Aguilar R 2017 *Constr Build Mater* **139** 193–202
[18] Aguilar R,Nakamatsu J, Ramírez E, Elgegren M, Ayarza J, Kim S and Ortega-San-Martin L 2016 *Constr Build Mater* **114** 625-637
[19] Danso H, Martinson D B, Ali M & Williams J 2015 *Constr Build Mater* **83** 314-319
[20] Khoshnevis B 2004 *Automat Constr* **13** (1) 5-19
[21] Buswell R A, Soar R C, Gibb A G and Thorpe A 2007 *Automat Constr* **16** (2) 224-231
[22] Cesaretti G, Dini E, De Kestelier X, Colla V and Pambaguian L 2014 *Acta Astronaut* **93** 430-450
[23] Gosselin C, Duballet R, Roux P, Gaudillière N, Dirrenberger Jand Morel P 2016 *Mater Design* **100** 102-109
[24] Perrot A, Rangeard D and Courteille E 2018 *Constr Build Mater* **172** 670-676
[25] Chiusoli A 2018 *Massa Lombarda: 3DWasp*Available: https://www.3dwasp.com/en/3d-printed-house-gaia/
[26] AENORUNE-EN ISO 17892-42019
[27] Lecompte T and Perrot A 2017 *Cement Concrete Res* **92** 92–97
[28] AENORUNE-EN ISO 17892-12 2019
[29] Mohanty A K, Misra M A and Hinrichsen G I 2000 *Macromol Mater Eng* **276 (**1) 1-24
[30] Li Y, Mai Y Wand Ye L 2000 *Compos Sci Technol* **60** (11) 2037-55
[31] Le T T, Austin S A, Lim S, Buswell R A, Gibb A G and Thorpe T 2012 *Mater Struct* **45** (8) 1221-32
[32] ASTM International ASTM D2573/D2573M 2018
[33] UNE 41410 2008 *AENOR*
[34] Wangler T, Lloret E, Reiter L, Hack N, Gramazio F, Kohler M and Flatt R *RILEM Technical Letters* **1** 67-75
[35] MehtaP K 1986 *Concrete: Structure, properties and materials* (New Jersey: Englewood Cliffs)
[36] Knudsen T 1984 *Cement Concrete Res* **14**(5) 622–630

IOP Conf. Series: Materials Science and Engineering **706** (2019) 012016 doi:10.1088/1757-899X/706/1/012016

Evaluation of fire, high-temperature and water erosion resistance of fiber-reinforced lightweight pozzolana-based geopolymer mortars

G Silva[1], J Salirrosas[1], G Ruiz[2], S Kim[1], J Nakamatsu[3] and R Aguilar[1]

[1]Engineering Department, Pontificia Universidad Católica del Perú PUCP, Av. Universitaria 1801, Lima 32, Lima, Peru

[2]Department of Civil Engineering, University of Piura, Av. Ramón Mugica 131, Urb. San Eduardo, Piura 353, Peru

[3]Science Department, Pontificia Universidad Católica del Perú PUCP, Av. Universitaria 1801, Lima 32, Lima, Peru

Email: raguilar@pucp.pe

Abstract. The development of new building materials requires intensive analyses of their durability under aggressive conditions. Therefore, the current study evaluated the capacity of a fiber-reinforced lightweight pozzolana-based geopolymer mortar to resist the action of fire, high temperatures and water erosion. The fire resistance and heat absorption capacity were evaluated through the exposure of disc samples to a direct flame at 1000 °C and the measurement of reverse-side temperature, respectively. To evaluate the high-temperature resistance, uniaxial compression tests were performed on cubic samples after their exposure to oven-temperatures of 300°C, 500°C, 800°C, and 1000°C. Finally, water jet erosion and permeability tests were performed to analyze the interaction of the material with water. The results of the fire tests indicate that fiber-reinforced lightweight pozzolana-based geopolymer mortars are able to absorb up to approximately 65% of heat flame with only slight degradation. On the other hand, uniaxial compression tests showed a temperature resistance up to 800°C, above this temperature the cubic samples experimented significant volumetric changes that might have been caused by a reaction of remaining hydrogen peroxide or gas expansion. Finally, water exposure tests showed that the material is highly permeable and has a remarkable resistance to water erosion.

1. Introduction

The ordinary Portland cement (OPC)-based concrete is one of the most used materials in the construction industry with worldwide average consumption of 3 tons per person each year [1]. The advantage of this building material is its versatility: it is used to produce a great number of elements for different applications and structures with many colors, densities, strengths and durability capacities. Also, in order to reduce the dead load of the structures and be easily manageable, low-density OPC-based concretes are developed by the addition of lightweight aggregates or by using foaming agents that increase their volume [2-4]. However, OPC production is one of the major contaminant industrial activities, due to its fabrication represents 5-7% of carbon dioxide emissions worldwide [2, 5-7]. For this reason, there has been greater development of new alternative materials to reduce the use of OPC leading to a more environmentally friendly construction industry.

Recent researches show that geopolymers are materials with similar properties in terms of mechanical properties and durability to OPC-based materials but with lower contaminant emissions in their production [6, 8, 9]. Geopolymers are obtained from the reaction between aluminosilicate sources and an alkaline activating solution. The most used aluminosilicate sources are industrial by-products and non-metallic minerals as fly ash and metakaolin and their mix with other waste materials [9]. In Peru, there is an abundant presence of natural pozzolana, a non-metallic mineral with high silica and alumina content that is extracted from volcanic ash deposits. In this context, a recent article has shown the possibility of using this material to produce high-strength geopolymers using natural pozzolana [11]. To produce lightweight geopolymer mortars a foaming agent was added to the mixture reducing the density. The presence of a foaming agent increases the volume of the material and generates a porous structure, known for its high degree of heat and water absorption [12, 13]. Additionally, 10-mm single jute fibers were added to the mixture to control shrinkage cracking problems. This article focuses on an evaluation of the durability of a fiber-reinforced light weight pozzolana-based geopolymer mortar under aggressive conditions by analyzing the effect of the action of fire, high temperatures and water erosion.

2. Materials and methods

2.1. Raw Materials

The materials used in this study consisted of a natural pozzolana and granulated quartz M40 as a precursor of the geopolymer and fine aggregate respectively, both materials provided by Compañia Minera Agregados Calcareos S.A (COMACSA).The natural pozzolana was extracted from Arequipa, Peru, while the granulated quartz M40 is a commercial product of the company. Particular size analyzer SediGraph 5100 was used to determine the particular size of both materials. The particle size analysis indicated that the natural pozzolana presented a d50 of 10.2 μm while the granular quartz M40 a d50 of 480.8 μm. The oxide composition of both materials is shown in Table 1.

Table 1. Oxidecomposition of natural pozzolan and granular quartz M40

Raw Material	Oxide							
	SiO_2	Al_2O_3	Fe_2O_3	K_2O	MgO	CaO	Na_2O	Others
Natural Pozzolan	53.55	10.81	0.98	4.07	0.55	1.43	2.30	26.31
Granulated Quartz M40	97.80	0.77	0.02	-	-	-	-	1.41

2.2. Alkaline solution and foaming agent

The alkaline solution used in this study was prepared mixing sodium silicate (28% SiO_2 + 8% Na_2O + 64% H_2O, technical grade, Abastecimientos Químicos Ciatex S.A.C.), sodium hydroxide in pearls (99.27% purity, technical grade, New China Chemicals Co., Ltd.) and distilled water. The dosage used in the preparation of the alkaline solution was SiO_2/Na_2O molar ratio of 1.08, Na_2O content of 8% and a water-binder ratio of 0.52. The foaming agent used was hydrogen peroxide (H_2O_2, 30% concentration, Movilab del Perú S.A.C.) with a content of 0.5% respect to the weight of the alkaline solution.

2.3. Jute Fiber

Jute is a plant fiber (*Corchoruscapsularis, Corchorusolitorius*) and mainly consists of cellulose (61-71%), hemicellulose (13.6-20.4%) and lignin (12-13%) [14]. Its usage in the geopolymer composite was decided based on its environmental benefits as Food and Agriculture Organization of the United

Nations (FAO) considers it as future fiber [15]. The jute fiber was obtained from commercial ropes and cuts to a length of approximately 10 mm.

2.4. Methods

The production of the fiber-reinforced lightweight pozzolan-based geopolymer mortar started by dry mixing pozzolana, fine aggregate (pozzolana to fine aggregate weight ratio of 2)and jute fibers (fiber content of 1% by weight of the pozzolana) for 1 min. Later, 95% of the alkaline solution was poured into the dry mixture and mixed for 3 min. The remaining 5% of the alkaline solution was combined with the hydrogen peroxide, then it was added into the previous mixture and mixed for another 1.5 min. Afterward, the lightweight mortar was placed into the corresponding molds for each test. Finally, they were cured in an oven at 65°C for 7 days.

First, the fire resistance was evaluated by the method proposed by Cheng and Chiu [16].This test was performed by applying a flame at 1000°C(Figure 1a) on the surface of a geopolymer disc of 100 mm diameter and 10 mm thickness for 30 min (Figure 1b) and measuring the temperature on the opposite side (reverse-side temperature).The temperature was measured by BTMETER BT-1500 Pyrometer 30:1 Laser Thermometer Gun with a precision of 0.1°C. Temperature measurement was performed until it reached room temperature (22 °C).The reverse-side temperature is an indicator of the degree of heat absorption capacity. The evaluation of this test was performed on three samples in two stages: Stage I evaluated disc heating while Stage II evaluated the cooling phase.

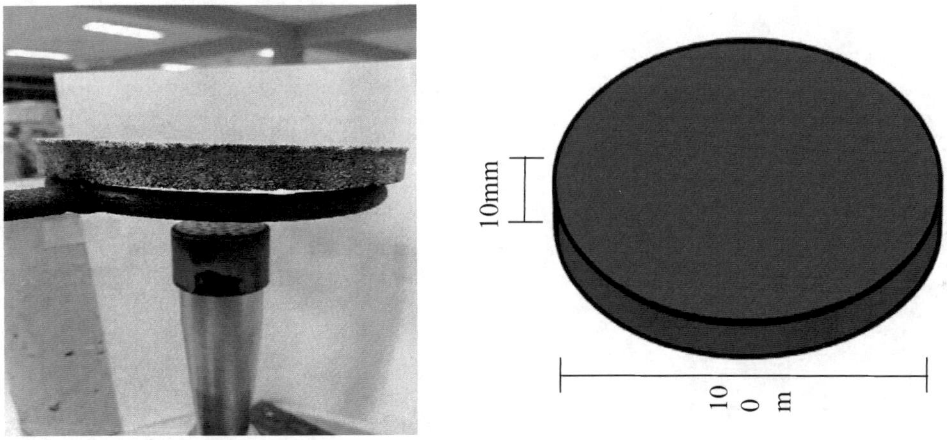

(a) (b)

Figure 1. Fire resistance test: (a) Test; (b) Disc dimensions.

The high-temperature resistance was measured by the method proposed by Ranjbar et al. [17] and RILEMTC 129-MHT [18]. Cubes samples (five per each temperature) were placed in an oven at 300, 500, 800 and 1000° Cusing a heating rate of 10°C per minute. After that, uniaxial compressive strength was determined following the guidelines of ASTM C109 [19], and the bulk density was measured using the Archimedes method. The results of the bulk density and compressive strength of heated samples were compared to the ones exhibited by samples left at room temperature (22 °C).

Water erosion was studied based on permeability [20] and water jet erosion tests [21]. The permeability was analyzed by measuring the contact angle using a Ramé-hart contact angle telescope-goniometer (Figure 2a) [22]. The test consisted of placing a drop of 10 μL of water on the surface of the fiber-reinforced lightweight geopolymer mortar samples (50 x 25x 10 mm prisms) and measuring the contact angle (Figure 2b). Low and high impermeability materials are classified based on the contact angle: permeable if the contact angle is less than 90° and impermeable if the contact angle is

IOP Conf. Series: Materials Science and Engineering **706** (2019) 012016 doi:10.1088/1757-899X/706/1/012016

greater than 90°. On the other hand, the water jet erosion test was performed following the methodology proposed by Benites et al. [21]. The test consisted of applying a water jet with a pressure of 200 psi from a distance of 300 mm (Fig. 2c) on one surface of a fiber-reinforced lightweight geopolymer mortar sample (300 x 200 x 100 mm plates) during 1, 5, 10 and 15 min. The surface condition after the water jet indicates the level of erosion caused by water exposure at high pressures.

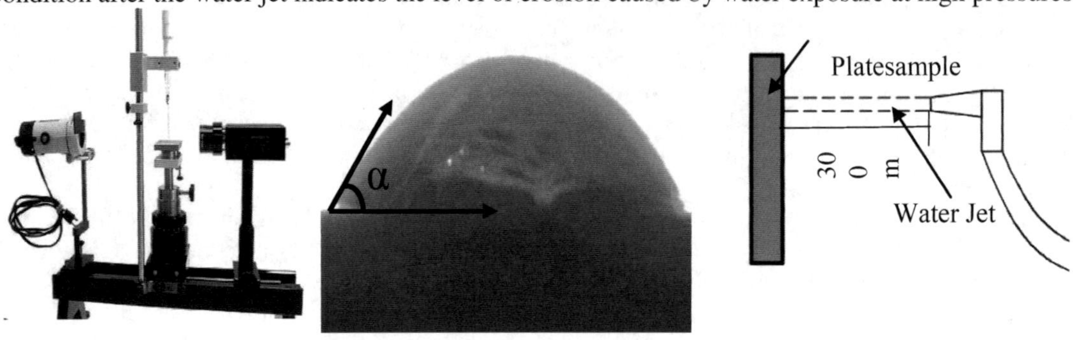

(a) (b) (c)

Figure 2. Water erosion test: (a) Ramé-hart contact angle telescope-goniometer; (b) Contact angle measurement; (c) Scheme of water jet test.

3. Results and discussion

3.1. Fire Resistance

Figure 3 shows the reverse-side temperature record during the fire resistance tests. At the beginning of Stage I, the material underwent a rapid change in its temperature until it reached its maximum temperature (less than 350 °C) and remained constant until the flame was extinguished. This significant retention of heat transmission can be explained because the unconnected pores trap the hot air avoiding the heat flow from the flame to the reverse-side surface. While in Stage II, the record of the reverse-side temperature indicates that the disc has a slight delay in cooling compared to heating. Nonetheless, cooling phase temperature has a greater uniformity than heating and the data fits an exponential decay of the temperature.

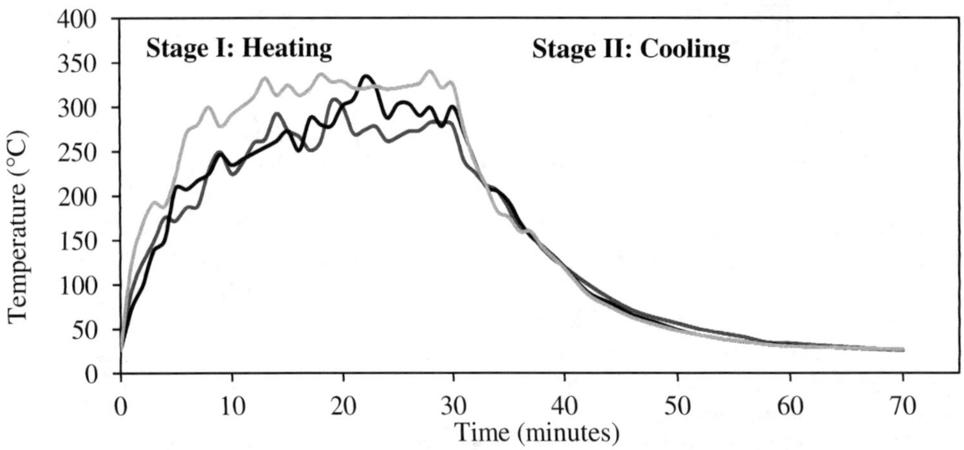

Figure 3. Reverse-side temperature vs. time from fire resistance tests.

As shown in Figure 4, the discs subjected to the action of fire retained their shape without noticeable deformations. However, there is a change of color to a darker one compared to the initial sample. This change of color was due to that the surface had a baking action by the effect of contact with fire. Another fact that can be deduced from the image is that the fibers added in the production still present in the sample, showing that the lightweight geopolymer acted as insulating protection to the action of fire.

(a) (b)

Figure 4. Fiber-reinforced lightweight geopolymer disc: (a) Before test; (b) After test.

3.2. *High-temperature resistance*

Figure 5 shows the specimens subjected to high temperatures. A change to a darker color was seen as the temperature increased with no significant volumetric changes up to a temperature of 800 °C. This change of color was due to the fact that the surfaces of the samples suffered a baking action by irradiated heat in the oven. This demonstrates that the fiber-reinforced lightweight-based geopolymer mortar is not severely affected by long-term exposure at temperatures up to 800 °C. However, in the case of specimens exposed to 1000°C, a significant loss of the initial shape was observed. Figure 5e shows a cross-section view of a cube sample exposed to an oven-temperature of 1000°C. An increase in the size of the pores caused by a new expansion process is observed.

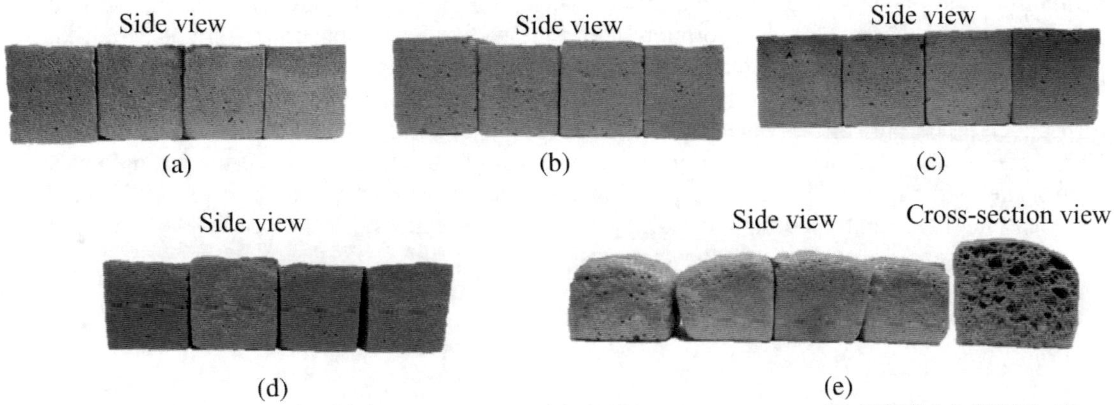

(d) (e)

Figure 5. Cubes exposed to high temperatures: (a) Ambient temperature; (b) 300°C;(c) 500°C; (d) 800°C; (e) 1000°C.

The mechanical and physical characterization of the cubes is shown in Figure 6. The results of the uniaxial compression tests (Figure 6a) show that the compressive strength of the fiber-reinforced lightweight geopolymer mortar does not change significantly up to 800 °C, considering the standard deviations. However, remarkable lower resistance to compression loads was obtained in samples exposed to 1000°C (4.65 MPa).On the other hand, it can be seen in Figure 6b that the density of the samples remains constant until a temperature of 500°C while the lowest density (0.66 g/cm^3) was recorded again at 1000°C. Similarly to qualitative analysis, mechanical and physical evaluation indicate that the fiber-reinforced lightweight geopolymer mortar is stable until 800°C.The reason why density does not dramatically change up to 800°C can be explained because there is no additional release of free water from the sample since it was released during the long curing period (7 days at 65°C). It is appreciable that the dramatic volumetric changes due to pore expansion results in a significant decreased in both compressive strength and density at 1000 °C

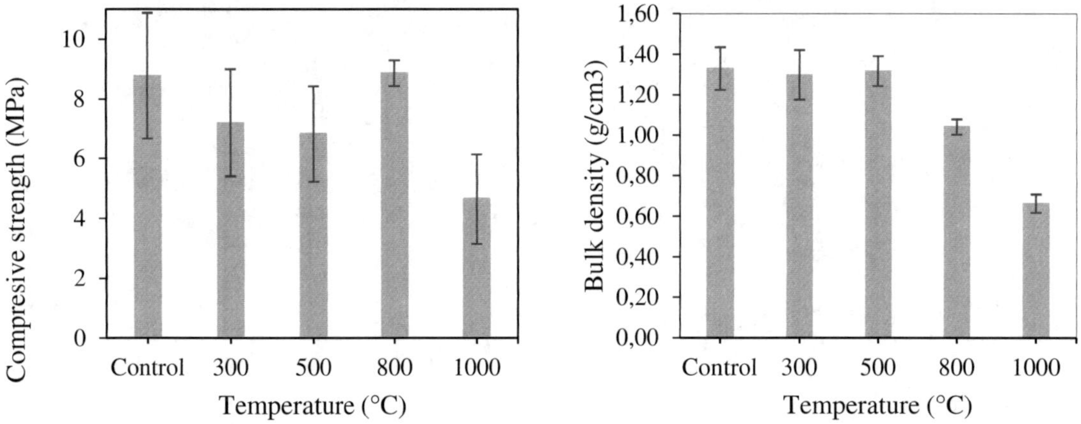

(a) (b)

Figure 6. Mechanical and physical characterization of fiber-reinforced light weight geopolymer mortar exposed to high temperatures: (a) Compressive strength; (b) Bulk density.

3.3. Water erosion tests

The contact angle tests indicated a contact angle of 43.31°± 9.06°. Therefore, the fiber-reinforced lightweight pozzolana-based geopolymer was classified as a material with low impermeability. In addition, it could be observed that the drops of water used for the test were absorbed in 3 seconds on average, demonstrating the high absorption rate of the material. Low impermeability is a well-known property of porous materials that tend to wet easily compared to denser ones.

Figure 7 a and b show the surfaces of the samples before and after the water jet erosion test, respectively. The pressurized water jet did not cause penetration into the surface of the specimens. Nonetheless, there is a slight erosion on the surface caused by the pressure of the jet and splashing water. The marks have a diameter of approximately 1.5 cm. Although it was proved in the previous test that the fiber-reinforced light weight geopolymer mortar is highly permeable, this material demonstrated remarkable resistance to water erosion.

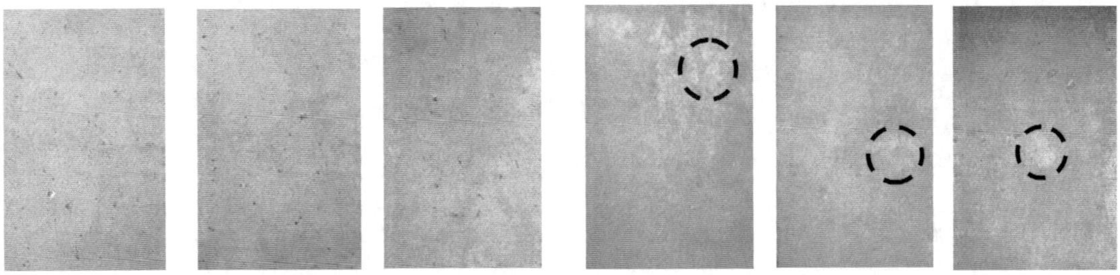

(a) (b)

Figure 7. Sample surfaces subjected to water jet erosion tests: (a) Before testing; (b) After testing.

4. Conclusions

This research focused on the effect on physical and mechanical characteristics caused by the exposure to aggressive conditions of specimens produced with a fiber-reinforced lightweight pozzolana-based geopolymer mortar. The fire-resistance test demonstrated that exist a considerable heat absorption capacity of the fiber-reinforced light weight pozzolana-based geopolymer mortar. The temperature recorded on the reverse-side surface did not exceed 350°C, which would be equivalent to a heat retention capacity of more than 65% of the heat irradiated by the flame. It was observed that the specimens resisted without a major change in shape with only a slight change of color to a darker one. In addition, the results also indicated appropriate insulating conditions to the layers inside the element. The exposure to high temperatures indicated that the specimens that were exposed to temperatures of 300°C and 500°C did not evidence significant changes in both compressive strength and density. However, at temperatures of 800°C, the density dropped significantly, but the compressive strength and shape did not significantly change. The most significant change occurred at a temperature of 1000°C due to a new process of pore expansion, the initial shape is lost and both properties, density and compressive strength, decrease considerably. Finally, fiber-reinforced lightweight pozzolana-based geopolymer mortar demonstrated to be highly permeable based on contact angle tests. However, the waterjet test showed great erosion resistance of the material to pressurized water with no suffering significant surface damage.

Acknowledgments

This work was supported by CONCYTEC and SENCICO PERU under the project: ‘‘GeoBloque: Desarrollo de bloques de construcción ultraligeros con geopolímeros” (Contract No. 105-207-FONDECYT). The authors are very grateful to Compañía Minera Agregados Calcáreos S.A (COMACSA) for providing raw materials. Finally, Guido Silva acknowledges the support from CONCYTEC (Contract N° 10-2018-FONDECYT/BM) for his Ph.D. fellowship funding.

References

[1] Van Oss H G 2019 *Mineral Commodity Summaries*

[2] Pimraksa K, Chindaprasirt P, Rungchet A, Sagoe-Crentsil K and Sato T 2011. *Mat SciEng: A* **528** (21) 6616–23

[3] Posi P, Teerachanwit C, Tanutong C, Limkamoltip S, Lertnimoolchai S, Sata V and Chindaprasirt P 2013 *Mater Design (1980-2015)***52** 580–586

[4] Hajimohammadi A, Ngo T, Mendis P, Nguyen T, Kashani A and Van Deventer J S 2017 *Mater Design* **130** 381–391

[5] Hardjito D and Rangan B 2005 *Curting University of Technology*

[6] McLellan B C, Williams R P, Lay J, Van Riessen A and Corder G D 2011 *J Clean Prod* **19** 1080-90

[7] Hardjito D, Wallah S E, Sumajouw D M J and Rangan B V 2004 *Australian J StructEng* **6** (1) 77–86
[8] Zhao R and Sanjayan J G 2014 *Mag Concrete Res* **63** (3) 163-173
[9] Moon J, Bae S, Celik K, Yoon S, Kim K, Kim K S and Monteiro P J 2014 *Cement Concrete Comp* **53** 97–104
[10] Yankwa Djobo J N, Elimbi A, Kouamo Tchakouté H and Kumar S 2016 *ConstrBuild Mater* **124** 606–614
[11] Silva G, Castañeda D, Kim S, Castañeda A, Bertolotti B, Ortega-San-Martin L, Nakamatsu J and Aguilar R 2019 *ConstrBuild Mater* **215** 633–643
[12] Hwang C and Hung M 2005 *ConstrBuild Mater* **19** (8) 619–626
[13] Haque M, Al-Khaiat H and Kayali O 2004 *Cement Concrete Comp* **26** (4) 307–314
[14] Yan L, Kasal B and Huang L 2016 *Comp Part B-Eng* **92** 94–132
[15] Food and Agriculture Organization of the United Nations (FAO) 2010 Future Fibres
[16] Cheng T and Chiu J 2003 *Miner Eng* **16** (3) 205–210
[17] Ranjbar N, Mehrali M, Alengaram U J, Metselaar H S C and Jumaat M Z 2014 *Constr Build Mater* **65** 114–121
[18] RILEM TC 129-MHT 2000 *Mater Struct* **33** 6-13
[19] ASTM InternationalASTM C109/C109M 2016
[20] Yuan y and Lee R 2013 *Surf. Sci. Tech.* **51** 3–34
[21] Benites V, Sánchez J and Ruiz G 2018 *Proceedings of the 16th LACCEI International Multi-Conference for Engineering, Education, and Technology: "Innovation in Education and Inclusion"*
[22] Aguilar R, Nakamatsu J, Ramírez E, Elgegren M, Ayarza J, Kim S, M. Pando A and Ortega-San-Martin L 2014 *Constr Build Mater* **114** 625-637

IOP Conf. Series: Materials Science and Engineering **706** (2019) 012017 doi:10.1088/1757-899X/706/1/012017

Neutral geopolymer foams reinforced with cellulose studied with the FT-Raman spectroscopy

M Szechyńska-Hebda[1,2], J Marczyk[3], C Ziejewska[3], N Hordyńska[1], J Mikuła[3] and M Hebda[3]

[1] The *Franciszek Górski* Institute of Plant Physiology Polish Academy of Sciences, Niezapominajek 21, 30-239, Cracow, Poland
[2] The Plant Breeding and Acclimatization Institute - National Research Institute, Radzików, 05-870 Błonie, Poland
[3] Institute of Materials Engineering, Faculty of Materials Engineering and Physics, Cracow University of Technology, Warszawska 24, 31-155 Cracow, Poland

E-mail: m.szechynska@ifr-pan.edu.pl

Abstract. In this work, cellulose fibers were successfully used as a reinforcement agent in a metakaolin-based geopolymer composite to produce, by direct foaming, sustainable and lightweight material with neutral properties for eco-friendly agricultural purposes. The structure of these materials after the neutralization process as well as changes during curing were investigated in their relations to the Raman spectra. The results showed both, a strong degree of conversion of the starting materials to the innovative neutral geopolymer product, and an advantage of theRaman spectroscopy in order to confirm the type of SiO_4 polymerization.

1. Introduction

Generally, polymer components used for industrial purposes are mostly derived from petroleum. Among others, plastics are produced most commonly from petrochemicals. Due to their low cost, ease of production, and perfect mechanical and physical properties, plastics are used in a multitude of products and for different industries, including agricultural purpose e.g. production of seed trays and pots. The statistic depicts the global plastic production increase from 1.5 million metric tons in 1950 to 348 million metric tons in 2017 [1]. We have calculated that the next generation (within 33 years) will produce 12 000–13 000 Mt of plastic, while the yearly consumption will reach 37–40 kilos of plastic per person worldwide [2]. Currently, a minimum of 40 percent of products are considered single-use plastic e.g. many pots are simply discarded after plants have been planted in the ground. The recent environmental reports have underscored the enormity of the global plastic waste issue, including microplastics, which abundantly appear in the ocean environment, soil, and air; they are ingested by fish;consumed by livestock, and thus included into the food chain;and finally, influence humans health [3,34]. It results in one of our planet's greatest environmental challenges. The current trends favor the development of a bio-economy and create demand for the use of more innovative and more renewable feedstock to ensure environmental protection. In the earlier study, we found that low-cost geopolymer foam materials could be a suitable alternative to plastic for agricultural purposes, provided they have neutral properties [2]. The excellent potential of a two-stage neutralization method was demonstrated, which allowed the use of geopolymers for the production of pots with approximately pH 7, and thus, their application for crop plant cultivation in ecological systems. Additionally, geopolymers foam pots were reinforced with cellulose fibres. This has improved the geopolymer structure, and its chemical, physical, and mechanical properties after the neutralisation process. The cellulose resulted in better adsorption and slower desorption of minerals in geopolymer foam pores during the fertilization and prevented elements from being washed-out during watering. This was crucial in the case of long-term fertilization effect, as geopolymer foams pots reinforced with cellulose fibres have distributed the absorbed macro- and microelements in a more uniform way and efficiently supplied them to plants. Consequently, the positive impact on plant growth was obtained when plants grew in the geopolymer foam pots compared to plants that had been grown in plastic containers [2].The reinforcement of the geopolymer with natural fibres such as cotton, sisal, and coconut is also an efficient method for improving mechanical properties e.g. fracture toughness and ductility[5].In the present study, the pots formed from basic geopolymer foams and geopolymer foams reinforced with cellulose, that were neutralised using two-step protocol, were used for investigation of the internal structure, its relation to the Raman spectra, and finally to evaluate the mechanisms responsible for better mechanical properties of cellulose-reinforced geopolymer composite.

2. Materials and methods

Ultrafine metakaolin powder KM 60 (Keramost) with chemical composition: 50–55% SiO_2, min. 40% Al_2O_3, 1.45% Fe_2O_3, 0.05–0.5% CaO, 0.20–0.45% MgO, 1.5% K_2O + Na_2; and quartz sand with chemical composition: 90.0–90.3% SiO_2, max. 0.2% Fe_2O_3, 0.08–0.1% TiO_2, 0.4–0.7% Al_2O_3, 0.17% CaO, 0.01% MgO, were used as base materials (in proportion 9:1). The process of alkaline activation was carried out according to [2]. Briefly, sodium silicate (R-145, a molar module of 2.5) wasmixed up with an 8M NaOH solution in a proportion of 2.5:1. In order to obtain foamed geopolymers, 99.7% purity aluminum powder (Fluca) in an amount of about 0.6 wt%, and 3.5% hydrogen peroxide solution (Sigma-Aldrich) in an amount of about 2wt% were added. Geopolymers were reinforced with cellulose ProMC2000 (ProAgro) with the length of cellulose fibers being 1350 μm, 50–70 g/L apparent density, and pH 7.0–8.0. The basic geopolymer foam pots (G) and the organic-geopolymer hybrid foams (GC) were treated twice with 0.5 M HCl solution, and the two treatments were separated by rinsing with distilled water.

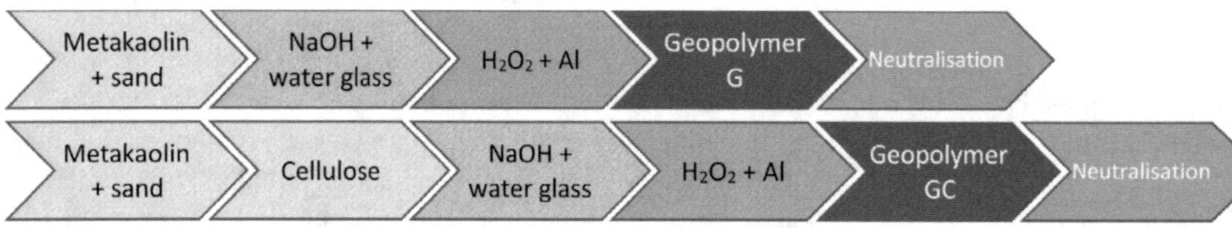

Figure 1. Steps of geopolymer production and composition of geopolymer materials.

The morphology of the materials was analyzed48h after sample formation with scanning electron microscopy JEOL-JSM-820 (JEOL, Tokyo, Japan). The samples were dried to a constant mass, stuck on the coal tape, and coated with a thin layer of gold using the JEOL JEE-4X vacuum evaporator (JEOL, Tokyo, Japan) to ensure good conductivity of the samples. The crystalline structure and mineralogical phases present in the sample were determined using X-ray diffraction analyses (Rigaku X-ray diffractometer) applying the following settings: CuKα radiation, a reflective graphite monochromator, 45 kV lamp voltage, 200 A lamp current, 0.05 2θ step and 1 s counting time per step. For phase identification, measured values of the lattice parameters were analyzed against the ICDD (International Centre for Diffraction Data 2016) catalog in the Xrayan program. Spectroscopy analyses were carried out for samples cured 48h or 3 months with an FT-Raman Nicolet NXR 9650 (Spectro-Lab) equipped with a Nd:YAG3 + laser with a 1064 nm wavelength on flat geopolymer samples after 48 hours and 3 months of curing at room temperature. For each material, a minimum of 3 spots were measured.The analyzed surface area of the sample was around 50 μm^2, while the depth of the focus was smaller than 1 μm. The Raman spectra were recorded in the region of 200–4000 cm^{-1}.

3. Results and discussion

Geopolymer is formed by activating aluminosilicate based-materials such as metakaolin with an alkaline solution, thereby creating structures of AlO_4 and SiO_4 molecules that get connected through oxygen atoms [6]. The product can be described by various degrees of polymerized SiO_4 tetrahedra [7]. The hydrogen peroxide solution is used as pore-forming agent, and induce the formation of a macroporosity and ultra-macro porosity with pore size > 50 nm and > 100 μm respectively.Following the reaction $2H_2O_2 \rightarrow 2H_2O + O_2$,hydrogen peroxide reacts, evolving oxygen gas, which is entrapped in the geopolymer paste[2, 8]. In our study, the porous structures were heterogeneous and pores with different size were randomly distributed in both, the basic geopolymer foam pots and geopolymer foam pots reinforced with cellulose. The most macropores were in the size range of 0.2–1 mm, however cavities up to 4 mm wide occurred also in the geopolymer network (Figure 2). The basic geopolymer foam (G) had higher number of macropores. Strong integration of the geopolymer foam with cellulose (GC) and its valid reinforcement resulted in reduced pore number. In the geopolymer gel matrix, fibres exhibited a random arrangement.

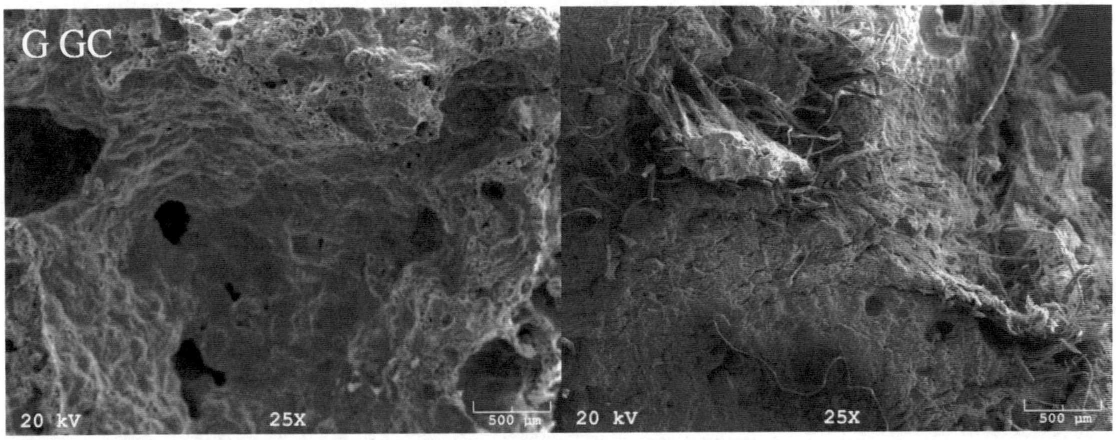

Figure 2. The representative microstructure of the basic geopolymer foam G and the organic-geopolymers hybrid foams GC after the two-step neutralisation procedure.

The presence of the crystalline phases originallyencountered in the metakaolin [9] i.e. quartz (SiO_2), mullite ($3Al_2O3_2SiO_2/2Al_2O_3SiO_2$), kaolinite ($Al_2Si_2O_5(OH)$, and illite $(K,H_3O)(Al,Mg,Fe)_2(Si,Al)_4\ O_{10}[(OH)_2,(H_2O)]$ was found in the X-ray diffractograms of the geopolymer foams.Thebasic geopolymer foams G showed a significant reduction in mullite and illite after the neutralisation process when compared to the content of mullite and illite in the geopolymers before acidic treatment(Table 1). Both mullite and illite are rich in aluminium (Al > Si), so their reduction was correlated with the dealumination process [2]. Geopolymers enriched with organic fibres GC exhibited a reduction of mullite and illite to a small extent. This proved that using cotton fibres can mitigate brittle failure in geopolymers and protect from the negative effect of acidic agents (during neutralisation).

Table 1. Changes in quartz, mullite, kaolinite, and illite after the neutralization process of the geopolymer foam pots, presented as the relative value of the phase (1.00 is fixed as the amount of component before neutralisation). White – an improvement in the phase after the neutralisation process in relation to its content before neutralization; grey – a reduction in the phase after the neutralisation process in relation to its content before neutralization.

	Quartz	Mullite	Kaolinite	Illite
G	1.09	0.49	2.22	0.72
GC	0.71	0.89	1.48	0.90

IOP Conf. Series: Materials Science and Engineering **706** (2019) 012017 doi:10.1088/1757-899X/706/1/012017

Applying Raman spectroscopy is a challenging task in the case of a geopolymer because this material exhibits a high level of fluorescence, which affects the spectra obtained. Using longer wavelengths can eliminate this problem. Using higher wavelengths, however, result in photoluminescence, which is most likely due to rare earth elements in the metakaolin [10, 11]. Another problem arises if the particles are too fine in geopolymer. Therefore analysis of geopolymer samples is rare in the literature. On the other hand, a SiO_4 tetrahedron, the building block of all silicates, amorphous or crystalline, is a very covalent entity and thus has a well-defined vibrational signature, especially in Raman scattering [7]. Furthermore, when the particular emphasis is placed with regard to the high-frequency regions, up to 4000 cm^{-1}, the stretching modes of OH bonds can be studied, which are relevant for the hydrated phases [12].

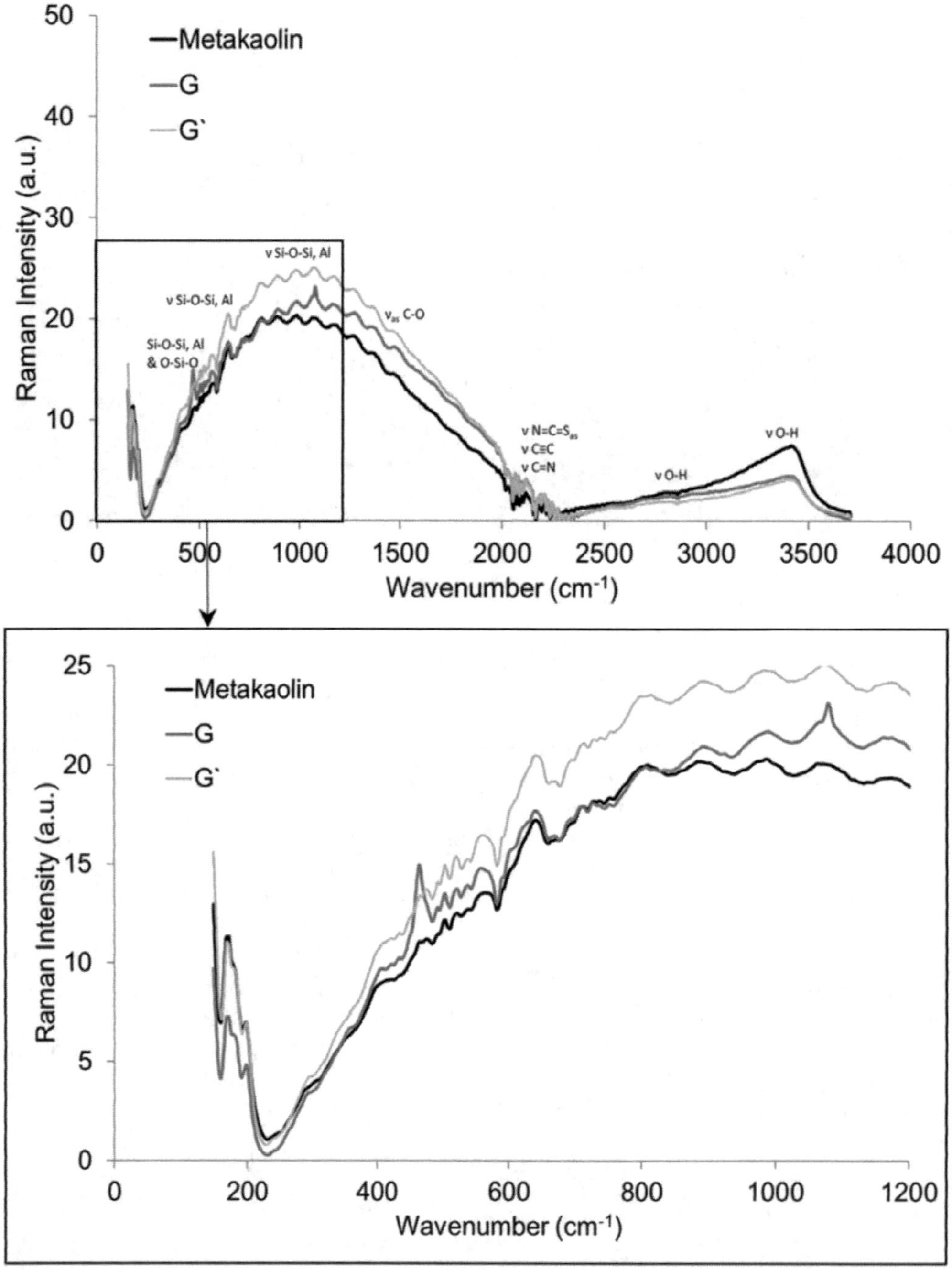

Figure 3. Raman spectra of the metakaolin and basic geopolymer foamsafter 48h (G) and 3 months of curing (G`).

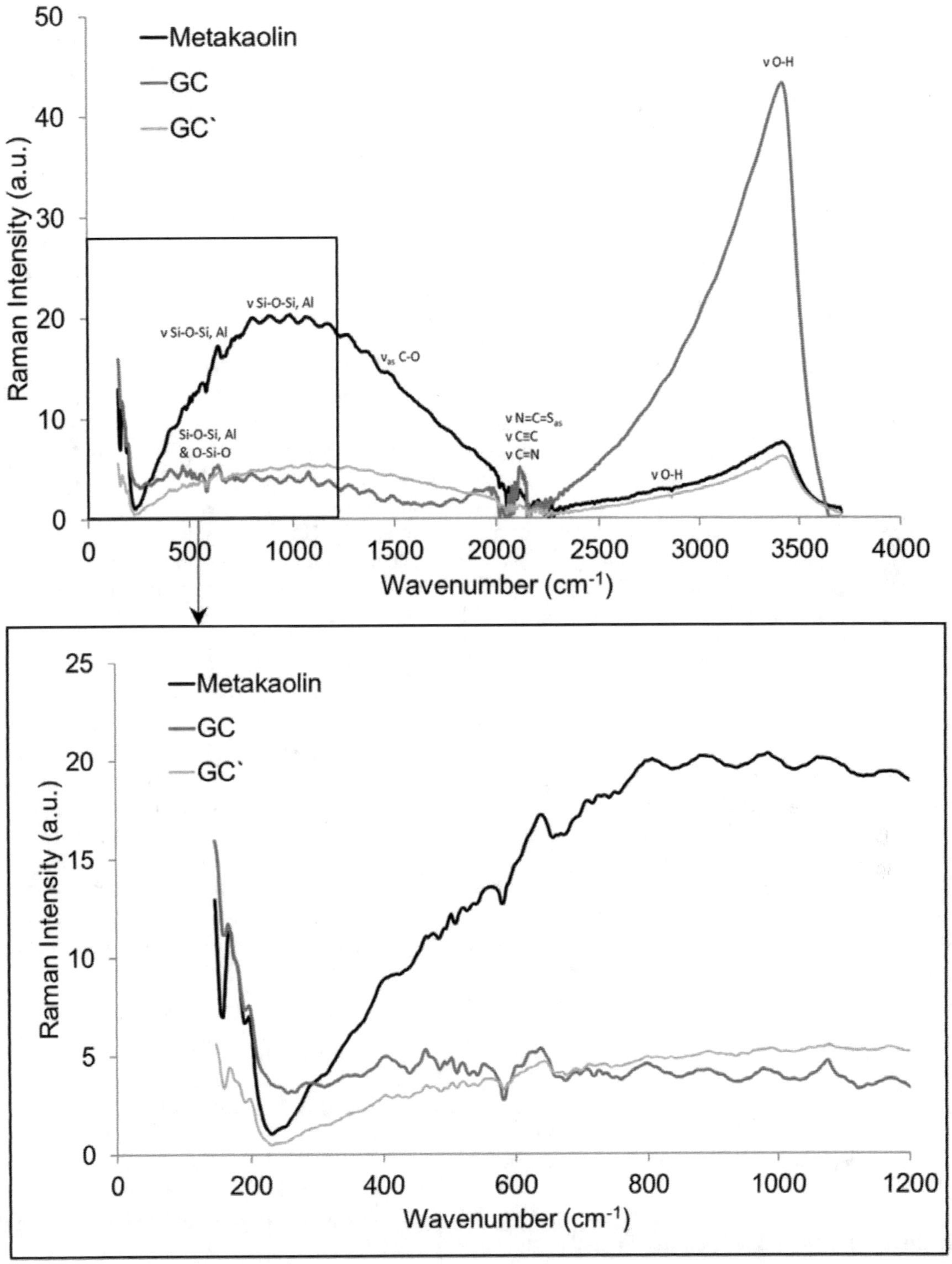

Figure 4. Raman spectra of the metakaolin and organic-geopolymers hybrid foams after 48h (GC) and 3 months of curing (GC`).

Figures 3 and 4 present the recorded spectra for geopolymer foam and organic-geopolymers hybrid foams cured at room temperature for 48 h and 3 months. Generally, the spectra of metakaolin, G and G' were similar, while GC and GC' differed significantly from metakaolin and from each other.

For basic geopolymers, multiple bondings in the 460–465 cm^{-1} range indicate the presence of Si–O–Si, Si–O–Al or O–Si–O bending vibrations, whereas in the 550–1159 cm^{-1} region, Si–O–Si or Si–

O Al symmetrical stretching vibrations were found. In addition, these materials exhibit weak effects at 3419 cm^{-1}, which correspond to O–H stretching vibrations. The intensity of the peaks for geopolymer reinforced with cellulose was significantly lower compared to metakaolin, particularly it concerns GC' samples, which were cured during 3 months. It can suggest a greater cross-linking of the aluminosilicate bonds and is in an agreement with our earlier studies which assumed that treatment of a geopolymer with a strong acid during neutralisation causes the breakage of the Si–O–Al bonds and dealumination of the geopolymers [2]. For G samples, the Si:Al ratio before neutralization was 11.94:12.06, while after neutralization the Si:Al was 29.94:3.24. Neutralization drastically reduced the Al content in G foams (37%), while the reduction for GC samples was to a lesser extent (15%) [2]. It suggests that chemical interaction between the organic and inorganic polymeric chains can prevent modification of the geopolymer structure in acidic conditions.Additionally, both mullite and illite are rich in aluminium (Al > Si), so their reduction demonstrated in Table 1 can be correlated with the dealumination process for the basic geopolymer, but not for geopolymers enriched with organic fibres GC. In the low frequency region of the spectrum, specific bands at 411, 459 cm^{-1}, 462 cm^{-1}, and 480 cm^{-1}, 916 cm^{-1}, 993 cm^{-1}, and 980 cm^{-1} can be assigned to intratetrahedral vibrations of polymerized SiO_4 tetrahedron. Some of the peaks from this group were absent or not intense in the spectrum of metakaolin. Further, it was suggested that broad bands near 430 cm^{-1} and 600 cm^{-1} are interconnected because they simultaneously increase or decrease in intensity [12]. Indeed, this effect was also observed for our spectra. These bands can attribute to starting components, originating from metakaolin, as higher peaks were observed for metakaolin and G, GC samples, in comparison to G', and GC' geopolymers. It can be concluded that these peaks describe thepolymerization effect, and indicate a formation of differently structured material after 3 months of curing. The peak at 430 cm^{-1} could be attribute to vitreous silica, while the peak centered near 600 cm^{-1} could be associated with broken Si-O-Si bonds or siloxane rings vibrations. The Raman bands in the region centered at 500 cm^{-1} attributed to the presence of bridging oxygens, or Si-O-Si and Si-O-Al linkages, and thus they are associated with tetrahedron polymerization. Similarly, a signal at 280 cm^{-1}, 412 cm^{-1}, and 522 cm^{-1},assigned to symmetric stretching (ν_1) of the Al–O bonds in the AlO_4^{5-} tetrahedrahas allowed to unambiguously distinguish the presence of this phase. Peak situated around 280 cm^{-1}, additionally can be ascribed to bonds in the $[(Fe,Al)O_4]^{5-}$ or $[(Fe,Al)O_6]^{9-}$ groups.All types of peaks, at 412 cm^{-1}, 500 cm^{-1}, and 522 cm^{-1}, were present in all geopolymer samples, howeverintensity of the peak at 412 cm^{-1} decreased in order: GC < G < GC' < G'.It indicates that the presence of this peak can be used to prove both, the effectiveness of geopolimerysation process and the cellulose effect in preventing dealumination during neutralization process. In the region corresponding to O–H stretching vibrations, a broad peaks between 3200 and 3800 cm^{-1}, emerge in the Raman spectrum. It was indicated that, these Raman spectra may be highly informative to identify the presence of hydrated aluminates in CAC and CSA [12].

4. Conclusions

Present studies confirm and broaden explanation of the mechanisms which occur during geopolimerysation process of materials with and without cellulose fibres. The better physical and mechanical properties of reinforced geopolymer composite resulted from an integration of fibres with geopolymer matrix, from the protective effect of cellulose fibre and inhibition dealumination during additional acidic treatment applied to lower pH of final product. Our research also pointed out that Raman spectroscopy is a suitable method to determine the degree and kinetics of polymerisation and curing processes. Our data confirm the high sensitivity of Raman spectroscopy to identify the main hydrated and unhydrated phases.

Acknowledgements

The authors are grateful to the National Centre for Research and Development, Poland for support from the project: "*Development of eco-friendly composite materials based on geopolymer matrix and reinforced with waste fibers*", under the ERANet-LAC 2nd Joint Call (http://www.eranet-lac.eu), program ERANet- LAC: Latin America, Caribbean and European Union funded by the European Commission, within the 7th Framework Programme for Research and Technology Development (FP7), project no. ELAC2015/T02- 0721. CZ, JM, JM and MH were financed from "*Development of*

3D printing technology for construction and facade prefabricated elements made of concrete composites and geopolymers" project no. POIR.04.01.04-00-0096/18.

References

[1] Garside M 2019 Global plastic production statistics – Statista, available online: https://www.statista.com/statistics/282732/global-production-of-plastics-since-1950/
[2] Szechyńska-Hebda M, Marczyk J, Ziejewska C, Hordyńska N, Mikuła J and Hebda M 2019 *Materials* **12** 2999; doi:10.3390/ma12182999
[3] Lebreton L and Andrady A 2019 *Palgrave Commun.* **5** (6) 1-11
[4] Will N and Niall S 2019 Verisk Maplecroft,available online: maplecroft.com (accessed on 2 July 2019,https://www.circularonline.co.uk/wp content/uploads/2019/07/Verisk_Maplecroft_Waste_Generation_Index_Overview_2019.pdf)
[5] Korniejenko K, Frączek E, Pytlak E and Adamski M 2016 *Procedia Engineering* 151 388-393
[6] Assaedi H, Alomayri T, Shaikh F and Low IM 2019 *Materials* **12**, 1459; doi:10.3390/ma12091459
[7] Kosora T, Nakić-Alfirevića B and Gajovićb A 2016 *Vib Spectrosc.* **85** 104-111
[8] Murri N, Medri V, Papa E, Laghi L, Mingazzini C and Landi E 2017 *Environments* **4** 86; doi:10.3390/environments4040086
[9] Mierzwiński D, Łach M, Hebda M, Walter J, Szechyńska-Hebda M and Mikuła J 2019 *J Therm Anal Calorim* https://doi.org/10.1007/s10973-019-08471-7
[10] Höhn S, Frimmel H E and Pašava J 2014 *Miner Deposita* **49** 967-986, https://doi.org/10.1007/s00126-014-0542-3
[11] Garg N, Wang K, and Martin S W 2013 *Cem. Concr. Res.* **53** 91–103 https://doi.org/10.1016/j.cemconres.2013.06.009
[12] Torréns-Martín D, Fernández-Carrasco L, Martínez-Ramírez S, Ibáñez J, Artús L and Matschei T 2013 *J. Am. Ceram. Soc.* **96** 11 3589-95

Utilization of innovative system for coke oven wastewater treatment as an element of stabilization technology for post-process waste from municipal incineration plants

D Mierzwiński[1], M Łach[1], J Mikuła[1], G Furtos[2] and K Korniejenko[1]

[1] Institute of Materials Engineering, Faculty of Materials Engineering and Physics, Cracow University of Technology, Jana Pawła II 37, 31-864 Cracow, Poland.
[2]RalucaRipan Institute of Research in Chemistry, Babes-Bolyai University, Str. Mihail Kogălniceanu, no 1 400084, Cluj-Napoca, Romania.

E-mail: dariusz.mierzwinski@mech.pk.edu.pl

Abstract. The article discuss the possibility of using the system for coke oven wastewater treatment process that has been developed within the framework of the project: *The innovative system for coke oven wastewater treatment and water recovery with the use of clean technologies — INNOWATREAT* that has received funding from the Research Fund for Coal and Steel under Grant Agreement No. 710078 and from the Ministry of Science and Higher Education Poland from financial resources on science in 2016–2019, for the removing the hazardous elements and compounds from wastewater from leaching ashes from municipal waste incineration plants. The results achieved in the project are compared with analysis provided for two different sources of ash from Poland and one Lithuania. The article discuss predicted effectiveness of the method and possibility their use in wider method of post-process waste stabilization technology as well as potential risk connected with implementation this solution into practice.

1. Introduction

Contemporary, there are about 540 coke plants in the world [1, 2]. The largest number is in Asia, where the leader is China – about 400 plants, but also the significant amount is in Europe – almost 60 plants [2, 3]. Among the European countries Poland is one of the biggest coke maker– table 1.

The production of the coke is connected with large requirements for water. It, after the industrial process, is one of the most contaminated and toxic aqueous stream of wastes [4, 5]. The coke wastewater contains organic and also inorganic pollutants [2, 4]. Moreover, their amount is significant - between 0.3 to 4 m^3 per ton of coke [1, 5].

Table 1. The top 10 of coke producers in 2014 - 2017 [3].

No.	Country	Coke oven, coke production (thousand metric tons)			
		2014	2015	2016	2017
1.	China	479,809	448,225	449,115	N/A
2.	Russia	39,512	40,342	26,236	39,934
3.	Japan	33,340	32,005	32,689	32,200

4.	South Korea	17,008	17,521	15,924	16,073
5.	Ukraine	13,858	11,617	12,722	N/A
6.	India	22,290	30,071	12,503	N/A
7.	United States	13,747	12,478	10,755	11,746
8.	Brazil	9,496	9,079	9,233	10,359
9.	Poland	9,568	9,792	9,718	9,431
10.	Germany	8,770	8,800	9,387	8,843

It is worth to notice that the problem of highly contaminated water is not only connected with coke making industry. The hazardous elements and compounds are also present in waste from others industry such as energy production, especially form coal. Producing and using coal affects the environment not only the air quality trough the gas emissions but also trough the combustion residues such as fly ash (FA) and bottom ash [6, 7]. The solid products of coal combustion such as coarse bottom ash and fine FA are usually between 5 and 20 wt.% of the feed coal [6, 8].The similar situation is for the municipal waste incineration plants, where instead of coal the solid wastes are burned. In this case bottom ash is between 15 and 25% by mass of the feed material and about 80–90% of the total residues [9].

The most of the residues form burning process, form coal as well as from municipal waste, consists of oxides of Si, Al, Fe, Ca and Ti (95-99%) and about 0.5% to 3.5% consists of Na, P, K, Mg, Mn and S. The rest of the FA is composed of trace elements such as As, B, Ba, Br, Cd, Cl, Co, Cu, Cr, Fe, Ga, Hg, I, In, Mo, Ni, Pb, Po, Rb, Sb, Sc, Se, Sr, Ti, W, V and Zn [10, 11], among them also toxic elements such as: Cr, Pb, Hg, Ni, V, As, and Ba. The presence of hazardous elements is more likely when we use as a feed municipal waste than coal. The concentration of the toxic elements in the FA is from 4 to 10 times higher than in feed material [6, 10, 12]. Because of that the most of FAs form municipal waste incrimination requires pre-treatment before the further application. One of the most popular is leaching. As a results of this process we received FA that could be used for further production process in many areas [13, 14] and contaminated wastewater [15].

The article compare the waste water form coke production and the results of leaching test for three FA - two different sources of FA from Poland and one Lithuania. The article discuss possibility of use the technology dedicated for treatment the wastewater form coke production to the treatment of post-process waste.

2. Technology

2.1. Technology - overall view

The new technology of the wastewater treatment has been developed in the framework of the project: *The innovative system for coke oven wastewater treatment and water recovery with the use of clean technologies - INNOWATREAT*. The project was implemented between 1 July 2016 and 30 June 2019. Project consortium include the partners with wide experience and knowledge on the field of coke making, coke oven wastewater characteristics and wastewater treatment and utilization methods, such as:

- Institute for Chemical Processing of Coal (Poland) - leader,
- Wroclaw University of Technology (Poland),
- Czech Technical University in Prague (Czech Republic),
- Cracow University of Technology (Poland),
- Akvola Technologies (Germany).

The main goal of the project was the development of the complex system for coke oven wastewater treatment and utilization of hazardous liquid waste. The performed activities included:

- Coke oven wastewater characteristics.

- Development of AkvofloatTM technology for tars removal.
- The replacement of conventional coagulation with electrocoagulation process.
- The feasibility studies preparation using computer modelling.
- Application of low pressure driven membrane processes in coke oven wastewater treatment.
- Technological grade water recovery from biologically purified coke oven wastewater.

This project allowed to develop the some conception of the complex systems for coke oven wastewater characteristics, treatment and utilization. The information about particular systems are available by the database: http://www.ichpw.pl/en/innowatreat-database/ [16].

2.2. Development of the technologies– exemplary solutions

The developed systems are presents as the schemes – figure 1 [16]. They could be a part of larger innovative system for coke oven wastewater treatment and water recovery.

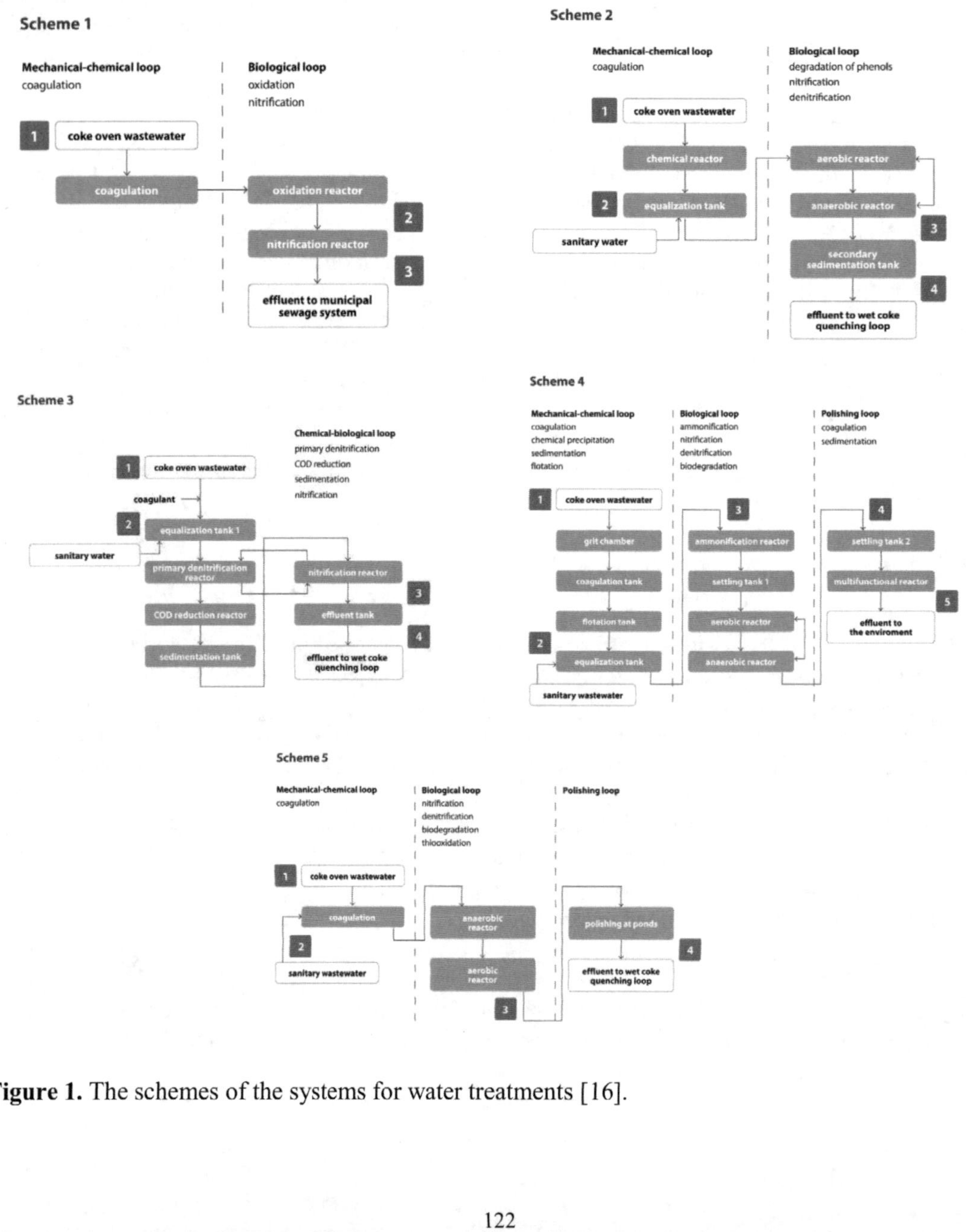

Figure 1. The schemes of the systems for water treatments [16].

The different systems are dedicate for different wastewater treatment technologies and have a little bit different efficiency.

2.3. *Effectiveness of the technologies for water treatment*

During the project INNOWATER the efficiency of different systems has been tested. The detailed description of the effects is presented in tables 2-6. The chemical analysis include the most characteristic elements and compounds was compared with permissible concentration in effluent for the best available technology (BAT). The numbers in table are connected with places marked on the schemes in the figure 1.

The table 2 presents the results for scheme 1. The final results (point 3) are worse than for BAT in case of amount of phenols and ammonia / ammonium nitrogen. The technology achieved better results in case of sulphides.

Table 2. Results of the wastewater treatment for the technology illustrated on scheme 1 [16].

Substance	Tar (sum 6 WWA) [mg/dm^3]	Cyanides [mg/dm^3]	Sulphides [mg/dm^3]	Phenols [mg/dm^3]	Ammonia/ Ammonium nitrogen [mg/dm^3]	Chemical oxygen demand [mgO$_2$/dm^3]
BAT permissible concentration in effluent	∑6 WWA: <0,05	(CN-) <0,1	<0,1	<0,5	total nitrogen: <50	COD<220
1 - coke oven wastewater	0,426	2,9-5,7 (4,3)	0,6-34,2 (17,4)	684-792 (738)	43-140 (92)	4101-5450 (4620)
2 - after oxidation	0,081	0,05-0,24 (0,15)	<0,05	7-13 (10)	24-105 (65)	344-375 (360)
3 - after nitrification	0,035	0,05-0,16 (0,11)	<0,05	4-6 (5)	15-60 (38)	288-363 (334)

The table 3 presents the results for scheme 2. The final results (point 4) are worse than for BAT in case of amount of phenols. The technology achieved better results in case of tar and sulphides.

Table 3. Results of the wastewater treatment for the technology illustrated on scheme 2 [16].

Substance	Tar (sum 6 WWA) [mg/dm^3]	Cyanides [mg/dm^3]	Sulphides [mg/dm^3]	Phenols [mg/dm^3]	Ammonia/ Ammonium nitrogen [mg/dm^3]	Chemical oxygen demand [mgO$_2$/dm^3]
BAT permissible concentration in effluent	∑6 WWA: <0,05	(CN-) <0,1	<0,1	<0,5	total nitrogen: <50	COD<220
1 - coke oven wastewater	0,17-3,38 (0,805)	0,4-25 (14)	<0,05	642-982 (772)	25-70 (37)	2740-4746 (3407)
2 - equalization tank	0,09-0,07 (0,027)	0,5-11 (7)	<0,05	404-534 (469)	15-56 (32)	1750-2436 (2199)
3 - after biological loop	0-0,022 (0,014)	0,05-1,5 (0,5)	<0,05	3-82 (20)	12-72 (58)	233-513 (398)
4 - water to coke wet quenching loop supply	0-0,022 (0,014)	0,05-1,5 (0,5)	<0,05	3-82 (20)	12-72 (58)	233-513 (398)

The table 4 presents the results for scheme 3. The final results (point 4) are much worse than for BAT in case of amount of phenols and ammonia / ammonium nitrogen. The technology achieved better results in case of tar and sulphides.

Table 4. Results of the wastewater treatment for the technology illustrated on scheme 3 [16].

Substance	Tar (sum 6 WWA) [mg/dm^3]	Cyanides [mg/dm^3]	Sulphides [mg/dm^3]	Phenols [mg/dm^3]	Ammonia/ Ammonium nitrogen [mg/dm^3]	Chemical oxygen demand [mgO$_2$/dm^3]
BAT permissible concentration in effluent	∑6 WWA: <0,05	(CN-) <0,1	<0,1	<0,5	total nitrogen: <50	COD<220
1 - coke oven wastewater	0,28	13	0,4-2,4 (1,4)	632-1500 (1066)	50-1900 (975)	6185-7650 (6917)
2 - equalization tank	0,07	0,05-2,8 (1,4)	0,05-1,2 (0,6)	265	44-1300 (630)	2134-2810 (2472)
3 - after nitrification	0,044-0,298 (0,171)	0,05-0,65 (0,35)	<0,05	15	38-1220 (478)	572-2029 (1290)
4 - water to coke wet quenching loop supply	0-0,046 (0,023)	0,05-0,5 (0,28)	<0,05	11-12,5 (11,7)	36-1140 (478)	378-1621 (1117)

The table 5 presents the results for scheme 4. The final results (point 5) are worse than for BAT in case of amount of phenols. The technology achieved much better results in case of tar, cyanides and ammonia / ammonium nitrogen sulphides.

Table 5. Results of the wastewater treatment for the technology illustrated on scheme 4 [16].

Substance	Tar (sum 6 WWA) [mg/dm^3]	Cyanides [mg/dm^3]	Sulphides [mg/dm^3]	Phenols [mg/dm^3]	Ammonia/ Ammonium nitrogen [mg/dm^3]	Chemical oxygen demand [mgO$_2$/dm^3]
BAT permissible concentration in effluent	∑6 WWA: <0,05	(CN-) <0,1	<0,1	<0,5	total nitrogen: <50	COD<220
1 - coke oven wastewater	0,021-0,051 (0,024)	0,3-15,8 (7,6)	<0,05	674-700 (686)	60-125 (87)	3057-4384 (3858)
2 - after mechanical-chemical loop	0-0,023 (0,012)	0,3-2,4 (1,6)	<0,05	610-632 (621)	60-125 (87)	3048-3622 (3335)
3 - after equalization tank	<0,01	0,19-27 (1,47)	<0,05	179		
4 - after biological loop	<0,01	0,05-0,5 (0,28)	<0,05	1,1-6 (3,6)	<0,5	220-428 (323)
5 - after polishing loop	<0,01	0,05-0,6 (0,3)	<0,05	0,5-4 (2,2)	<0,5	201-335 (252)

Table 6. Results of the wastewater treatment for the technology illustrated on scheme 5 [16].

Substance	Tar (sum 6 WWA) [mg/dm^3]	Cyanides [mg/dm^3]	Sulphides [mg/dm^3]	Phenols [mg/dm^3]	Ammonia/ Ammonium nitrogen [mg/dm^3]
BAT permissible concentration in effluent	∑6 WWA: <0,05	(CN-) <0,1	<0,1	<0,5	total nitrogen: <50
1 - coke oven wastewater	0,2	16,7	6,6	884	106

2 - after coagulation	0,165	10,6	1,2	859	122
3 - after biological loop	0,163	<0,05	<0,05	4	92
4 - water to coke wet quenching loop supply	0,009	<0,05	<0,05	2	

The table 6 presents the results for scheme 5. The final results (point 4) are much worse than for BAT in case of amount of phenols. The technology achieved better results in case of tar, cyanides and sulphides. The information for and ammonia / ammonium nitrogen is not given.

Taking into consideration the amount of the contamination in the coke oven wastewater the achieved results are satisfactory.

3. Leaching results for fly ashes and discussion

The leaching test was carried out using 3 types of waste from municipal waste incineration plants: one FA proceeding from a waste incineration plant in the city of Klaipeda in Lithuania - FA1, and two FAs from a waste incineration plant in Konin (Poland) - FA2 (fly ash containing hazardous substances) and FA3 (boiler dust containing hazardous substances). In all cases the pre-treatment of waste was made by the same method - the washing process. The FA was rinsed with running water in the ratio of 1:5 in a slow-running laboratory mixer (time: 15 minutes), then the excess water was filtered and dried in a dryer to a constant mass.

The samples for the leaching test were prepared according to the PN-EN ISO 15587-2:2005 standard on the samples witch mass approximately 2.5 kg each. The tests were conducted according to the PN-EN 12457-4:2006 standard. The results are shown in table 7.

Table 7. The results of water leaching tests for FAs.

Analysed substance	Criteria for admission of hazardous waste	FA1	FA2	FA3
		mg/kg	mg/kg	mg/kg
Arsenic (As)	25	<0.5	<1.0	<0.2
Barium (Ba)	300	76.5	82	3
Cadmium (Cd)	5	0.25	<0.05	<0.01
Chromium (total amount) (Cr)	70	0.26	0.5	2.1
Copper (Cu)	100	225	<0.2	<0.05
Mercury (Hg)	2	<0.01	<0.01	<0.01
Molybdenum (Mo)	30	1.5	1.6	1.8
Nickel (Ni)	40	<0.05	<0.05	<0.05
Lead (Pb)	50	365	125	2.1
Antimony (Sb)	5	<0.5	<0.5	<0.2
Selene (Se)	7	<0.1	<0.5	<0.2
Zinc (Zn)	200	44.5	58	18.5
Chlorides (Cl^-)	25 000	162 000	177 000	15 000
Fluorides (F^-)	500	44	51	34
Sulphates (SO_4^{2-})	50 000	27 800	8 880	14 200
Dissolved organic carbon (DOC)	1 000	24	60	26
Total dissolved solids (TDS)	100 000	360 000	448 000	69 900
Chromium (VI)		<1.0	<1.0	<4.0

The achieved results was compared with the criteria for admission of hazardous waste. The FA3 do not exceed the regulations. The some problems appears in case of FA1 and FA2. For the FA1 the amount of cooper, lead and TDS is over the limit. On case of FA2 the values for lead and TDS are too high. In this two cases the wastes required further treatment and limitation of TDS up to 100,000 mg / kg.

The technologies of the wastewater treatment developed in the framework of the project: *The innovative system for coke oven wastewater treatment and water recovery with the use of clean technologies – INNOWATREAT* do not taken into consideration the compounds of cooper and lead, so there is impossible to investigate their efficiency for this two elements. The main question in this area is connected with possibilities of TDS removing.

TDS is the combination content of dissolved all inorganic and organic substances present in a liquid – wastewater [17, 18]. The particular chemical composition of TDS is unknown. It limited the speculation about effectiveness of the technology, but taking into consideration the wide range the pollutions that has been removed from coke wastewater it could be also possible to use this system for the treatment the waste water after leaching the FA. The using this system will required further investigation and probably optimization some procedures, because the effectiveness of this procedure in case of FAs is strongly depended on time [18, 19]. That is mean in practice that the treatment procedures required the proper time and it effectiveness will be limited by amount of wastes. Other problem could be the risk associated with changeability of wastes. The usefulness of implemented technology will be depended on the amount of contamination is the base material. It could limited the application the proposed solution in practice.

4. Conclusions

The content of heavy metals and toxic elements content in FAs is important factor than give a possibility of their safe utilization. The presented FAs were pre-treated by leaching because they exceeded the limits for hazardous waste landfills. In the case of one of them the procedure was effective, but for two of them it was not sufficient. In this case the more advanced treatment technologies are required. The article discussed the possibility of using the system for coke oven wastewater treatment process for the removing the hazardous elements and compounds from wastewater from leaching ashes from municipal waste incineration plants. The comparison achieved results show that is impossible to use this system without any changes and it required further research works to be applicable in wider process of post-process waste stabilization technology for municipal waste incineration plants.

Acknowledgements

1. This work has been supported within the framework of the project: *The innovative system for coke oven wastewater treatment and water recovery with the use of clean technologies — INNOWATREAT* that has received funding from the Research Fund for Coal and Steel under Grant Agreement No. 710078 and from The Ministry of Science and Higher Education Poland from financial resources on science in 2016–2019.
2. This work has been supported by Intelligent Development Operational Program 2014-2020, 1.1.1: Industrial research and development works carried out by enterprises, funded by the National Centre for Research and Development in Poland, within the framework of the grant: 'Development and implementation of eco-innovative post-process waste stabilization technology' ('Opracowanieiwdrożenieekoinnowacyjnejtechnologiistabilizacjiodpadówpoprocesowych'), grant no. POIR 01.01.01-00-0740 / 16-00.

References

[1] Rychlewska K, Kwiecińska A, Kochel M and Figa J 2018 *Desalin. Water Treat.* **128** 207
[2] Ksepko E, Klimontko J and Kwiecinska A 2019 *J. Therm. Anal. Calorim.*. https://doi.org/10.1007/s10973-019-08214-8

[3] Database FactfFsh 2019 *Coke oven coke, production (thousand metric tons) - for all countries*published on line: http://www.factfish.com/statistic-country/ukraine/coke%20oven%20coke%2C%20production (access: 18-09-2019)
[4] Kwiecińska A, Kochel M, Rychlewska K and Figa J 2018 *Desalin. Water Treat.* **128** 214
[5] Kwiecińska A, Figa J and Stelmach S 2016 *Pol. J. Environ. Stud.* **25** (2) 465
[6] Belviso C 2018 *Prog. Energ. Combust.* **65** 109
[7] Eliche-Quesada D, Sandalio-Pérez J A, Martínez-Martínez S, Pérez-Villarejo L and Sánchez-Soto P J 2018 *Ceram. Int.* **44** 4400
[8] Guzmán-Carrillo H R, Pérez J M, Aguilar Reyes E A and Romero M 2018 *Int. J. Environ. Sci. Technol.* **15** 1757
[9] Vavva C, Voutsas E and Magoulas K 2017 *Chemical Eng. Research & Design* **125** 57
[10] Yadav V K and Fulekar M H 2018 *Int. J. of Advance Eng. & Research Develop.* **5** (04) 768
[11] Roy W R, Thiery R G, Schuller R M and Suloway J J 1981 Coal fly ash: a review of the literature and proposed classification system with emphasis on environmental impact (Champaigm, Illinois: Environ. Geol. Notes 96)
[12] Bouaissi A, Li L Y, Moga L M, Sandu I G, Al Bakri Abdullah M M and Sandu A V 2018 *Rev. Chim. (Bucharest)* **69** (7) 1661
[13] Mucsi G, Kumar S, Csőke B, Kumar R, Molnár Z, Rácz A, Mádai F and Debreczeni Á 2015 *Int. J. Miner. Process.* **143** 50
[14] Łach M, Mikuła J, Hebda M 2018 *J. Therm. Anal. Calorim.* **125** (3) 1035
[15] Łach M, Mierzwiński D, Korniejenko K, Mikuła J and Hebda M 2018 *JAPCA J. Air Waste Ma.* **68** (11) 1190
[16] Institute for Chemical Processing of Coal 2019 *Innowatreat – database*published on line: http://www.ichpw.pl/en/innowatreat-database/ (access: 18-09-2019)
[17] Atanes E, Cuesta-García B, Nieto-Márquez A and Fernández-Martínez F 2019 *J. Environ. Manage.* **240** 359
[18] Haberl J and Schuster M 2019 *Waste Management* **87** 268
[19] Verma C and Verma R 2019 *Nature Environ. & Pollution Techn. Int. Quart. Sci. J.* **18** (2) 403

IOP Conf. Series: Materials Science and Engineering **706** (2019) 012019 doi:10.1088/1757-899X/706/1/012019

Possibilities of using the 3D printing process in the concrete and geopolymers application

J Marczyk[1],C Ziejewska[1], M Łach[1], K Korniejenko[1], W T Lin[2] and M Hebda[1]

[1] Institute of Materials Engineering, Faculty of Materials Engineering and Physics, Cracow University of Technology, Jana Pawła II 37, 31-864 Cracow, Poland.
[2] Department of Civil Engineering, National Ilan University, No. 1, Section 1, Shennong Road, Yilan City, Yilan County, 260 Taiwan.

E-mail: mhebda@pk.edu.pl

Abstract. Additive manufacturing, also known as three-dimension printing (3DP) has a lot of advantages over traditional construction technology, including high building efficient, less construction wastes and greatly reduced labor. This paper presents the possibility of applying this technique to 3D printing of geopolymers and concrete. Moreover, a general overview of the research and progress of 3D printing of geopolymers and concrete in recent times was characterized. This paper also identifies the most interesting and innovative solutions in this area and highlights emerging trends to stimulate further research.

1. Introduction

In the last few years, additive manufacturing commonly known as 3D printing is gaining more and more interest, especially in the construction industry [1-7]. This technology creates components directly from a digital model [7, 8]. An innovative process which is three-dimension printing technology can be used with wide applications in many areas, including aerospace and automotive [7-8], fabricating objects [9], concrete components [10], food [11] and medical treatments [12]. Additive manufacturing techniques have a significant asset in the production of components with complex structures and in terms of higher productivity, quality, cost-efficiency and with less waste material [2, 4, 7, 13].

Nowadays, 3D printing of construction materials poses new opportunities for researchers and engineers in this industry and can revolutionize the traditional building practices [1-2, 14-15]. However, the use of suitable cementitious material of 3D printing is critical to conducting a successful process [16].

One of the main problems that should be solved during the 3D concrete printing process is understanding how the material can be sufficiently fluid and at the same time have sufficient viscosity to retain its shape after the printing process [14, 17]. For this purpose, the concrete mix created must have a thixotropic character, and thus should have a low viscosity during flow and high yield strength at rest [14].Considering these challenges, scientists and industrialists have created cement materials that are easily extruded and retain their shape after the process [14, 17-19].

Unfortunately, despite the fact that technology is developing, the use of conventional raw materials, such as, ordinary Portland cement (OPC) causes direct release of CO_2 into the atmosphere, which has a negative impact on the environment [16-20]. Therefore, to strive for sustainability, one of the alternatives to Portland Cement may be the use of local materials [16]. One of the materials that use locally available products such as slag or fly ash as ingredients are geopolymer. It is well known that the term geopolymer includes a class of modern, amorphous, inorganic aluminosilicate polymers with specific composition and properties.Geopolymers are produced by thefollowing reactions: (1) Dissolution of metakaolin into silicate monomers and aluminate monomers;(2) polymerisation of monomers into aluminosilicate oligomers, and then into small geopolymerfragments or 'proto-zeolitic nuclei' (thermodynamically metastable and incompletely cross-linked); (3)fragments combination into larger molecules, that finally form aluminosilicate inorganic polymer gels,and crystallised phases, consisting of SiO_4 and AlO_4tetrahedra sharing oxygen corners [21, 22].Moreover, geopolymer material can achieve even better mechanical properties than ordinary Portland cement,and may be used in the construction industry [16, 23].

This paper presents the possibility of applying the 3D printing process for geopolymers and concrete applications. Moreover, the potentials of using this technology in terms of materials, mechanical properties and feasibility were presented. Finally, selected applications of 3D printed geopolymers and concrete were discussed.

2. Definition and principles of 3D printing of concrete

The development of concrete 3D printing results from interdisciplinary research of scientists in the fields of materials engineering, design, calculations, architecture and robotics [1, 24]. Because traditional technology and thus vibration and formwork setup are not required for 3D printing, printed concrete combines the advantages of self-compacting concrete and spray concrete [1].

Modeling of shapes for 3D printing must be carried out in accordance with sets of requirements related to processing restrictions, e.g. element dimensions, layer thickness, etc., and the functional properties of the manufactured part, e.g. mechanical strength. Functional requirements depend mainly on the structural geometry and properties of the hardened material [24].

The scheme of the 3D printing setup is shown in figure 1.

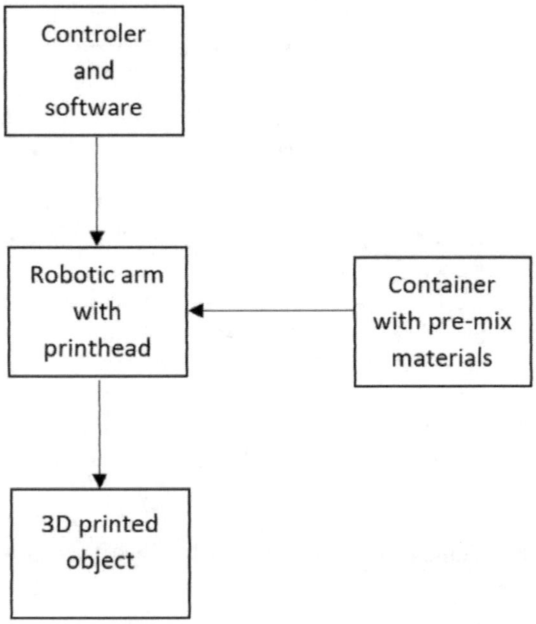

Figure 1. Schematic of the 3D printing setup [based on 24].

In order to implement 3D printing technology, the first step is to design a building path using 3D-to-2D slicing software. First, the constructor cuts the 3D shape of the object into thin layers of specified thickness, and then creates the path of each layer. All paths consist of a contour line and a fill pattern [21]. The basic principle of this process is that the nozzle moves along the marked path, after which the concrete is extruded. Construction is completed after laying all layers for printing [1, 24].

3. The current progress of 3D printing of concrete

Panda and Tan [6] created a novel 3D printing geopolymer mortar, which can be applied for printing non-structural building components. Moreover, in this solution, it can be directly printed from digital models without the need of any formwork. They tested their produced materials with methods for laboratory testing 3D printing concrete (figure 2).

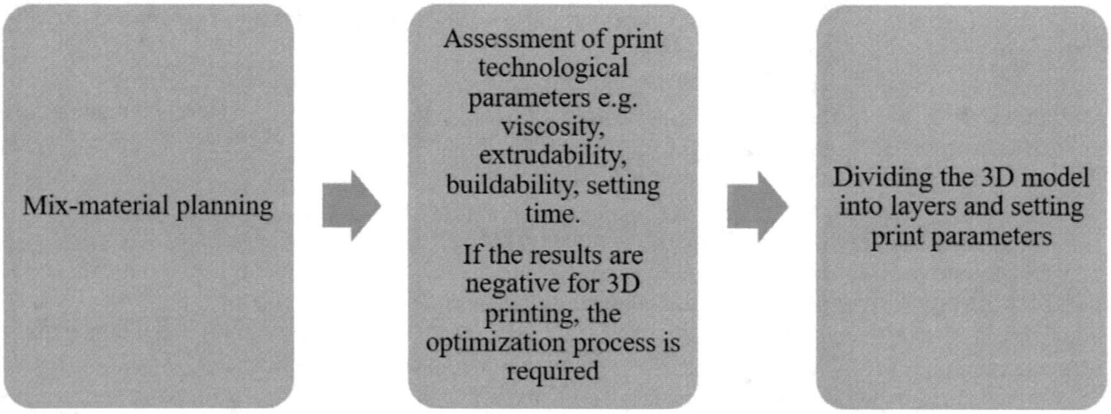

Figure 2. Diagram of an experimental method for research 3D printing material [based on 6].

The results of the conducted laboratory tests confirmed that produced fresh geopolymer has properties, which are necessary for 3D printing application. In addition authors evidence that proposed criteria and mix design are suitable in practice. The open 3D structure using the optimum mixture, which based on fly ash, by a 4-axis gantry concrete was printed. The total printing time for this element was less than 20 min with a flowrate of 0.5 l/min and a printing speed of 80 mm/sec. The height was 60 cm, whereas width was 35 cm. This component comprised of 10 mm thickness 60 layers [6].

Inherent problems with 3D printing concrete rely on low tensile strength and poor ductility due to non-reinforcement. These aspects greatly limit the application of 3D printing materials and structures. Ma and al. [25] formed a reinforced geopolymer composite by using a continuous micro steel cable (1.2 mm) during filaments (12 mm) deposition process.The composition of used geopolymer (in percentage of total weight) are F-class fly ash (0.64), silica fume (0.11), slag (0.25), silica sand with particle size ranging from 0.1 mm to 0.6 mm (1.2), sodium metasilicate pentahydrate powder (0.125), tap water (0.348) and also small dosage of polypropylene fibers (0.0056) to prevent shrinkage. Moreover to increase water-retention property in this mix also used appropriate viscosity modifying admixture (VMA, Hydroxyethyl cellulose). Figure3illustrates the schematic manufacturing of filament reinforced composite. The most important part of the printing process is e.g. the step motor-based active feed-system. An extruder repeatedly entrains the filament into the printing nozzle by a narrow transmit pipe. After that the filament is extruded equally with the deposition of geopolymer paste, forming a reinforced composite [25].

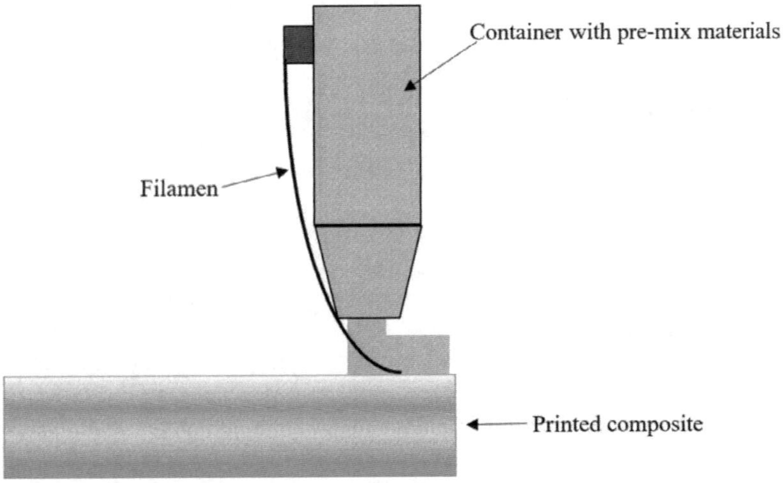

Figure 3. Schematic illustration of the 3D printing process of composite reinforced by a filament.

The results show that the 3D printed micro-cable reinforced geopolymer composite gains eight times higher value of flexural strength and seventy times higher deflection resistance than the non-reinforced one [25]. These performances were obtained when the filaments were deposited in an incline-crossed printing configuration.

Al-Qutaifi et al. [26] focused on studying some factors related to the impact on the structural buildability of layered objects. A special timber mold was prepared to simulate the layering process by extrusion. Three geopolymer mixtures were prepared and three-time gaps of 5, 10 and 15 minutes are used. Two-layered patterns were tested with 18 layered samples. In addition, the authors also assessed the impact of 3D printing on hardened mechanical properties of geopolymer concrete by using 9 control samples. Materials such as fly ash (class F), fine aggregate (sand), activators (the combination of 8M sodium hydroxide solution and D-grade sodium silicate) and reinforcing materials were used in the research. The weight ratio of sodium silicate to sodium hydroxide was 1, and the weight ratio of alkaline liquid to fly ash was 0.26. Mixes 1, 2 and 3 contained 0%, 1% steel fibers and 0.5% polypropylene fibers, respectively. Al-Qutaifi et al. examined the flexural strength of layered geopolymer elements. On samples carried out three 3-point bending tests for 27 samples after 12 days of curing at room temperature and 2 hours of curing in an oven at 70°C. The results show that through the use of 3D printing technology, the geopolymer-based material has adequate bending strength. Despite this, its strength is lower than after the casting process. This is due to the layering processes. The flexural strength of the geopolymer material decreases due to anisotropic phenomena resulting from the presence of empty spaces between successive layers. In contrast, the flexural strength can be improved by increasing the thickness of printed layers. Reducing the time intervals between layers also increases the flexural strength of the layered element. The authors found that the introduction of fibers into the material can reduce the bond strength between successive layers. This is due to the fact that the uneven distribution of fibers can hinder adhesion between layers. Steel fibers, compared with polypropylene fibers, had a higher decrease in bond strength. It is not recommended to use steel fibers as reinforcement during 3D printing, as this may hinder adhesion between successive layers.

Zhu et al. [27] carried out research aimed at developing 3D-printing engineered cementitious composites (ECCs) characterized by high tensile strain capacity. These composites, in principle, can be used for the digital construction of concrete structures, moving away from conventional methods that use steel fibers. In this study, the authors used cementitious binder materials that consisted of ordinary Portland cement, sulfoaluminate cement-based on 100% solid industrial waste and fly ash. Fine silica sand has also been used. However, to obtain the required fluidity, a superplasticizer based on polycarboxylate ether-based was introduced. Samples M1 and M2 were made to examine the effects of adding HPMC (hydroxypropyl methylcellulose) and ANC (attapulgitenanoclay). While in order to strengthen the ECC matrix, various volume fractions of polyethylene fibers (1%, 1.5%, and 2%) were introduced into samples M2, M3, and M4. For each of the mixtures M1 and M2, a cylindrical model was printed, which consisted of 17 layers, the height of each layer was 13 mm. Sample M2 showed excellent buildability and shape retention due to the addition of HPMC and ANC. Although the M1 sample was not damaged, deformations of the printed layers and poor shape retention could be observed.

Moreover, Zhu et al. [27] examined also rheological properties, workability, uniaxial tensile strength, flexural and compressive strength as well as microstructure for ECC developed for 3D printing. For comparative purposes, the samples were also made using a conventional casting method. The results showed that the tensile strength, as well as deformation energy and the number of cracks, increased with the volume fraction of PE fibers in the material. ECC tensile strength depended on the fiber distribution and pore size distribution. The extrusion-based 3D printing process caused the fiber orientation to be parallel to the direction of the force. The flexural strength of 3D printed materials was higher than that of mold-cast ECCs. However, the compressive strength of printed ECCs was lower than that of mold-cast ECCs. The impact of fiber content on compressive strength was small.strength of 3D printed materials was higher than that of mold-cast ECCs. However, the compressive strength of printed ECCs was lower than that of mold-cast ECCs. The impact of fiber content on compressive strength was small.

4. The most interesting applications of 3D printing of concrete

The application of 3D printed concrete has become really popular nowadays, thus some structural components and also full-size buildings can now be printed. Unfortunately, many printed structures and buildings can't be used, because of the unqualified safety requirements [1]. The first workable 3D printed building was created in Dubai in 2016. The cost of production this building was relatively modest - $140,000. This building is used as an office building. It was created with used a single $120 \times 40 \times 20$ feet 3D printer.

The first 3D printed pedestrian bridge was built in 2017 by the Institute of Advanced Architecture of Catalonia (IAAC) in Spain. It has 12 m span and 1.75 m wide and it consists of eight separate components [1]. Every single component was built in the D-Shape printing system. In addition, Salet et al. [28] printed a bridge by CC printing system, which is located in the Netherlands. The authors of the project divided the bridge into six components and they printed them apart. This was caused by the span limitation of the printer.

5. Summary

The article provides a general view of knowledge related to 3D printing of geopolymers and concrete. Moreover, the article presents problems that arise during the additive manufacturing process. 3D printing technology was also compared to conventional methods. It was determined what properties the material obtained depending on the method of manufacture.

It was presented that, the success of using 3D printing technology is based on improving the relationship between design and production processes, but it also depends on the skills of engineers to design building components. The quality of the printed elements is also undeniably an important value, which is determined by the material efficiency and precision of production. The printability of concrete or geopolymers depends on the mechanical properties and workability that can be optimized by selecting materials and printing parameters. The technology of 3D printing of concrete and geopolymers brings low labor costs, less waste material, and high efficiency. Currently, some buildings have already been printed, but 3D printing technology for concrete and geopolymers still requires continuous development and optimisation.

IOP Conf. Series: Materials Science and Engineering **706** (2019) 012019 doi:10.1088/1757-899X/706/1/012019

Acknowledgements
The authors are grateful to the National Centre for Research and Development, Poland for support from the projects: "*Development of 3D printing technology for construction and facade prefabricated elements made of concrete composites and geopolymers*" project no. POIR.04.01.04-00-0096/18.
The authors are grateful to the Polish National Agency for Academic Exchange for support under the International Academic Partnership Programme within the framework of the grant: *E-mobility and sustainable materials and technologies EMMAT*.

References
[1] Zhanga J, Wanga J, Donga S, Yu X and Han B 2019 *Compos Part A.* **125** 105533
[2] Li B, Weng Y, Li M, Qian Li B, Weng Y, Li M, Qian Y, Fai Leong K, Jen Tan M and Qian S 2019 *Constr Build Mater.* **207** 477–490
[3] Xia M andSanjayan J G 2018 *Mater Lett.* **227** 281–283
[4] Hui Lin J, Panda B and Pham Q-C 2018 *Constr Build Mater.* **178** 32–41
[5] Zhang D, Wang D, Lin X and Zhang T 2018 *Constr Build Mater.* **184** 575–580
[6] Panda B. and Tan M J 2018 *Ceram Int.* **44** 10258–265
[7] Xia M. and Sanjayan J 2016 *MaterDesign* **110** 382–390
[8] Xia M., Nematollahi B and Sanjayan J 2019 *Automat Constr.* **101** 179–189
[9] Fantino E, Chiappone A, Roppolo I, Manfredi D, Bongiovanni R, Pirri C F and Calignano F 2016 *Adv. Mater.* **28** 3712–17
[10] Buswell R A, Leal de Silva W R, Jones S Z and Dirrenberger J 2018 *Cement Concrete Res.* **112** 37–49
[11] Sun J, Zhou W B, Huang D J, Fuh J Y H and Hong G S 2015 *Food Bioprocess Technol.* **8** 1605-15
[12] Seol Y J, Kang H W, Lee S J, Atala A and Yoo J J 2014 *Eur. J. Cardiothorac. Surg.* **46** 342-348
[13] Lowke D, Dini E, Perrot A, Weger D, Gehlen C and Dillenburger B 2018 *Cement Concrete Res.* **112** 50-65
[14] Panda B, Unluer C and Tan M J 2018 *Cement Concrete Comp.* **94** 307-314
[15] Panda B, Unluer C and Tan M J 2019 *Compos Part B* **176** 107290
[16] Biranchi P, Suvash C P, Nisar A N M, Yi Wei D T and Ming J T 2018 *Measurement* **113** 108-116
[17] Suvash C P, Yi Wei D T, Biranchi P and Ming J T 2018 *Arch CivMechEng.* **18** 311-319
[18] Franchin G, Scanferla P, Zeffiro L, Elsayed H, Baliello A, Giacomello G, Pasetto M and Colombo P 2017 *JEurCeram Soc.* **37** 2481-89
[19] Le T T, Austin S A, Lim S, Buswell R A, Law R, Gibb A G F and Thorpe T 2012*Cement Concrete Res.* **42** 558-566
[20] Suhendro B 2014 *Procedia Engineer.* **95** 305-320
[21] Szechyńska-Hebda M, Marczyk J, Ziejewska C, Hordyńska N, Mikuła J and Hebda M 2019 *Materials* **12** 2999; doi:10.3390/ma12182999
[22] Mierzwiński D, Łach M, Hebda M, Walter J, Szechyńska-Hebda M and Mikuła J 2019 *J Therm Anal Calorim* https://doi.org/10.1007/s10973-019-08471-7
[23] Łach M, Mikuła J and Hebda M 2016 *J Therm Anal Calorim* **125** (3)1035–45
[24] Gosselin C, Duballet R, Roux Ph, GaudillièreN, Dirrenberger J and Morel Ph 2016 *Mater. Design* **100** 102-109
[25] Ma G, Li Z, Wang L and Bai G 2019*Mater Lett.* **235** 144-147
[26] Al-Qutaifi S, Nazari A and Bagheri 2018*Constr Build Mater.* **176** 690-699
[27] Zhu B, Pan J, Nematollahi B, Zhou Z, Zhang Y andSanjayan J 2019*MaterDesign***181** 108088
[28] SaletT, Ahmed Z, Bos F P andLaagland H L M 2018*Virtual Physical Prototyping* **13** 222-236

AUTHOR INDEX